Domestication

the decline of environmental appreciation

Domestication
the decline of environmental appreciation

HELMUT HEMMER

TRANSLATED INTO ENGLISH BY
NEIL BECKHAUS

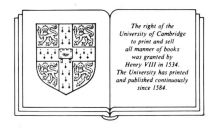

The right of the
University of Cambridge
to print and sell
all manner of books
was granted by
Henry VIII in 1534.
The University has printed
and published continuously
since 1584.

CAMBRIDGE UNIVERSITY PRESS

Cambridge

New York   *Port Chester*   *Melbourne*   *Sydney*

Flower Veterinary Library
September 14, 1990

Published by the Press Syndicate of the University of Cambridge
The Pitt Building, Trumpington Street, Cambridge CB2 1RP
40 West 20th Street, New York, NY 10011, USA
10 Stamford Road, Oakleigh, Melbourne 3166, Australia

First published in German 1983
Second edition in English 1990

Printed in Great Britain at the University Press, Cambridge

British Library cataloguing in publication data

Hemmer, Helmut
Domestication.
1. Livestock. Behaviour
I. Title
636

Library of Congress cataloguing in publication data

Hemmer, Helmut,
[Domestikation, English]
Domestication: the decline of environmental appreciation/Helmut
Hemmer: translated into English by Neil Beckhaus.
    p.   cm.
Translation of: Domestikation.
"First published in German, 1983"—T.p. verso.
Includes index.
ISBN 0–521–34178–7
1. Domestication.   2. Domestic animals.   I. Title.
SF41.H4613   1990
636—dc20 89–9993 CIP

ISBN 0 521 34178 7 English edition
(ISBN 3 528 085054 German edition)

Contents

Preface to the German edition

In order to understand biological systems, more than a formal description of their precise structure is required. Even a knowledge of their highly complex biochemical foundations or an insight into their biomechanical functioning is not sufficient. All this is undoubtedly a prerequisite for total comprehension, but in the end only an overall view provides a really profound understanding of the life of organisms. One has to attempt to construct a complete picture from the many small pieces of the mosaic provided by single disciplines in the biological sciences. The more pieces that are available, the more the rough outlines, which are all that is initially visible, can be refined and filled in so as to bring the full picture closer.

The study of domestic animals provides a good example of this general problem of advancing knowledge in the sciences by using painstaking studies of single problems from all branches of biological research, while constantly renewing attempts to combine the information available at any one time to form an up-to-date picture. The enormous importance attached to comprehending the 'domestic animal' as a special zoological phenomenon was appreciated by Charles Darwin in the last century. Nevertheless, research on domestic animals has been fragmentary. There have been few attempts to produce an overview. Those who do attempt overviews inevitably leave gaps and are criticized by the producers of fragments of information.

A uniform overall concept for the 'domestic animal' phenomenon may acquire practical significance over and above purely scientific interest in fundamental knowledge, since, as will be shown in the first chapter, the domestic animal is an essential element in the development of human civilization. An understanding of domestication as a developmental process under human control should be convertible into realistic strategies for planned production of quite distinct kinds of domestic animal.

Such strategies might improve food production for peoples living on the threshold of subsistence, build up alternative methods of land use, obtain new laboratory animals for use in urgent problems of medical research or gain new breeds which can be productive in today's intensive husbandry techniques without welfare problems. This book should provide an incentive and a foundation for such research. Using information from previous studies on domestic animals and from the recent work of the author's research group, it endeavours to outline, in an intelligible way, an overall concept of the phenomenon of the 'domestic animal' and to present a completely new view of this. The analysis has been intentionally restricted to the Mammalia, since this class is very diverse and contains most of the important domestic animals. Mammals are the subject of particularly wide interest, and all the essential problems can be studied using them.

The central point for understanding the phenomenon of the 'domestic animal' lies in the structure of its behaviour, especially in the underlying changes from the wild condition. In this connection, it has proved to be extremely fruitful to revive and extend the concept of environmental appreciation, based on the term *Merkwelt* or 'perceptual world', coined by the Baltic zoologist and founder of ecological research, Jakob von Uexküll. Environmental appreciation will be used here to designate the totality of perception and the evaluation of the components, characteristics and events in the environment of an animal. In other words, the world as the animal experiences it. There is a large span between an environmental appreciation which is impoverished because of genetic or environmental inadequacies, and one which is enriched. There is a network of relationships that links seemingly independent factors such as stress and psychosocial tolerance, behavioural flexibility, activity and intensity of action, aptitude for life in social groups and differentiation of social relationships, sexual and aggressive reactions, and even pigmentation and bodily development, into a closely interwoven system. That this network could also serve to provide a more profound knowledge of human nature independently of the 'domestic animal' phenomenon may be mentioned in passing. The parallels – and any ideas implicit in them – will not escape the attentive reader.

The gradual progress achieved by the author's research group, which is concerned with problems of mammalogy in general and the study of domestic animals in particular, played an important role in the origin of the overall concept presented here. Although the members of this group have changed several times over the years, the mutual goal of attaining understanding persisted and led to ever deeper insight. In view of the very different importance of individual contributions, a list of all co-workers who were involved in one way or another at any time is omitted here. Those responsible for providing important pieces of the mosaic are mentioned at the relevant places in the text. Thanks for the painstaking execution of the drawings used in this book are due to Ms Käthe Rehbinder. Finally the *Vieweg-Verlag* have kindly made possible the generous layout and so provided a form that will surely facilitate the understanding of the contents.

Mainz, August 1982   Helmut Hemmer

Preface to the English edition

Since the publication of the original German version of this book, many people from the international community involved in aspects of the life of domestic animals, from archaeozoologists to game farmers and zoo managers, have asked me for an English translation of the new synthetic view of the nature of domestication put forward. They would then have better access to the various aspects of interest for their own work. Greatly influential in having the translation published by the Cambridge University Press were Dr Juliet Clutton-Brock of the British Museum of Natural History, London, and Dr Colin P. Groves of the Australian National University, Canberra, to both of whom I proffer my best thanks for their really valuable help in successive stages of this work. I am also grateful to Robin Pellew, Susan Sternberg, Martin Walters, Robin Smith and Sandi Irvine, Cambridge University Press, for being so co-operative during the negotiations for the publication of the English edition. I was fortunate in being able to have the book translated by Neil Beckhaus of my home department at the Johannes Gutenberg University, Mainz, which allowed constant discussion and checking during the process of translation and considerably shortened the preparation time necessary.

The text has been updated for this edition by the inclusion of new research results, ideas and discussions. The most far-reaching relevant event in the years since the publication of the original German edition has been the successful and novel domestication of a large mammal by the strict application of the basic concept originally outlined as a strategy for domestication. The fallow deer project described in outline in Chapter 12 has finally crossed the borderline of the transition zone from the wild to the domestic animal within the second breeding generation and produced the first primitive, but truly domestic fallow deer. Details on the history and course of the project and subsequent experience gained during the dissemination of the new domestic deer will be published in the future.

Mainz, August 1988   Helmut Hemmer

1 | Why are domestic animals kept?

Domestic animals may be provisionally defined as those kept and bred in and around human habitation to be used constantly to human advantage. Such animals are found in flats in large modern cities, in the form of innumerable dogs, cats, rabbits, guinea pigs and hamsters. We find them on the farms of Europe where cattle, horses and donkeys, sheep and goats, pigs, dogs and cats, geese, ducks and hens are kept. Various combinations of species are to be found around the huts and villages in the tropical lands of the Old World. We see camels, horses, sheep and yaks amongst the tents and yurts of the herdsmen of Central Asia and large herds of reindeer amongst the people of the far north of Europe and Asia. In Europe and Asia, as amongst the Eskimos of the icy wastes of the frigid zone of the North American continent, the indigenous Polynesians of the Pacific Islands, the San of the Kalahari, or the Indians in the tropical forests of South America, the dog is the most widely distributed domestic animal. In short, domestic animals are to be found everywhere human beings have made their abode. They are our universal companions (Figs. 1.l–1.4).

Other animals that are kept, but not bred over generations for a particular use, nor subjected to any particular kind of selection, cannot be termed domestic animals, even in the widest sense. They are wild animals kept in captivity and sometimes tamed, as for example are small monkeys, tropical wild cats, pet birds, lizards or tortoises. As a rule, such animals do not undergo the process of domestication. This also holds in principle for zoo animals. It goes without saying that animals that live as

Fig. 1.1. Domestic animals are characteristic of farms all over the word. In addition to cattle and pigs, goats as 'the small man's cow' and donkeys as beasts of burden for many different purposes used to be a part of village life in Central Europe (farmhouse in Tessin, Switzerland).

1

unwelcome guests in human habitations, for example, mice, flies, mosquitoes, cockroaches and bugs, also have nothing to do with domestication.

Domestic animals have provided an essential basis for the fulfilment of the most elementary human needs since the Neolithic at the latest, so it is no surprise that the course of history as a whole has for millennia been closely connected with that of domestic animals. The domestication of the ruminants – sheep, goats, cattle and camels – was the prerequisite for the development of the nomadic herding way of life, producing conflict over land use with sedentary peasant economies from the very beginning. The Cain and Abel story of the Old Testament hints at this rivalry. From the standpoint of the herdsman's culture, within which that metaphorical story was first handed down, the embodiment of good in the herdsman

Fig. 1.2. The horse as a draught and saddle animal has contributed to immense historical upheavals. In the steppes from East Europe to Central Asia, where the origins of its domestication are to be sought, it still fulfils decisive functions in the day-to-day life of mounted herdsmen (rural horses near a yurt in Kirghizia).

Fig. 1.3. Pigs bred solely for meat production are just as much at home among huts in tropical villages as in countries in the temperate zones. They rank first in the world statistics of meat production (village in southwestern Madagascar).

Abel and of evil in the farmer Cain appears completely comprehensible. This tense relationship continues through the millennia right up to the conflicts between farmers and Aborigines in Australia and farmers and Indian tribes in South America. After the domestication of the horse, horsemen from the wide steppes of eastern Europe to eastern Central Asia time and again changed the course of history: the Hyksos succeeded in conquering the Middle Kingdom in Egypt around 1650 BC, the Huns increased the pressure on the peoples of Europe during the migratory movements as 'Antiquity' gave way to the Middle Ages, in the fourth and fifth centuries AD, and the Mongols under Genghis Khan created a huge empire from the east of Europe to the China Sea at the beginning of the thirteenth century. The horse also made it easier for the Spaniards to subjugate the great Inca and Aztec empires of Central and South America at the beginning of the sixteenth century.

The universal distribution of domestic animals and their significance in changing the course of history is based on the many benefits human society has been able to gain from them. This is best expressed by figures on the world stocks of some important species: over a 1 000 000 000 each of cattle and sheep, over 500 000 000 pigs, and almost that number of goats. One elementary need that domestic animals take an essential part in fulfilling is that for food (Figs. 1.5–1.7). Instead of acquiring meat by hunting, which ties up more manpower and involves a higher risk of failure, it gradually became more and more common to keep herds of domestic animals as a constant supply. Most kinds of domestic animal are, or at least were earlier in their history, used as fattened stock for meat production. In 1978, for instance, the worldwide production of meat amounted to about 132.4 million tonnes, of which 48.1 million tonnes were beef and 48.7 million tonnes pork. In addition, several large species of herbivores provide milk and milk products such as butter and cheese.

Fig. 1.4. The dog is the oldest and originally the most widely distributed domestic animal. There are dogs even in civilizations which have no other domestic animals (young dingo, Australian feral dog).

Milk production world-wide now amounts to over 450 million tonnes a year (Fig. 1.8). The Masai in East Africa use live cattle as a further source of food by regularly extracting blood from them. Eggs, an important product of domestic poultry, play just as large a part in nutrition as the milk of cattle and other large mammals.

A human necessity just as basic as food, at least in non-tropical regions, is clothing as protection against the cold. Domestic animals provide hides, which are processed into leather for the most varied articles of clothing from shoes to overcoats. Their coats are used for fur products; sheep and rabbits are the main sources, although dogs and cats are also

Fig. 1.5. In industrial countries, highly rationalized pig fattening in narrowly confined pens serves to guarantee meat supplies.

Fig. 1.6. Cattle occupy second place in the world's meat production, close behind pigs. Fattening young bulls is of great importance.

used, and animals such as foxes and minks, first farmed in the second half of the last century, have acquired considerable economic significance for their high quality fur. For instance, in 1976 West Germany imported 4.75 million mink pelts for DM242.9 million and 6.20 million sheep skins, mainly Persian lamb, for DM278.5 million. The third raw material significant for clothing production is wool, above all from sheep but also from goats and alpacas. In addition to the protection woollen clothing gives against cold, wool is also used in making carpets which insulate the floors in our homes. The production of wool in the world amounts to over 1.5 million tonnes a year.

People's requirements for warmth can be satisfied either directly or

Fig. 1.7. The black and white cattle originating in the lowlands of Holstein and Friesia are one of the best dairy breeds in the world, with an annual yield of over 5500 kg milk per cow. They have now become widely distributed.

Fig. 1.8. Rationalized milk production in industrialized countries presupposes large stocks of cattle. The extent of these stocks in a modern dairy is most striking when the cows crowd into their shed after coming from the pasture.

indirectly by domestic animals in other ways. Dingoes, the original Australian domestic dogs now found mostly in the feral state, still serve the Aboriginal people as living sources of warmth on cold desert nights when human and animal sleep closely huddled together. In several desert and dry areas the dried-out dung of camels, llamas or yaks can be used instead of wood as fuel while simultaneously giving light. Fat from domestic animals is also used in oil lamps; honeybees, which may for the present be termed domestic animals according to the generalized definition, provide wax for candles.

Domestic animals have been exploited in a great variety of ways for medical purposes. For instance, almost all parts of the dog's body were used in Graeco–Roman medicine. Dog fat, and later also cat meat, were supposed to help against tuberculosis. For centuries until the present day, cat skins have been used as an effective therapy for rheumatic diseases. Many parts of the goat's body were significant in folk medicine. Until genetic engineering and biotechnology provide higher yields, domestic animals are still indispensable as sources of several hormones, such as insulin for the treatment of diabetes or thyroxine to treat thyroid ailments. Domestic animals are used in the production of serums for vaccinations. Pigskin may be a temporary substitute skin in the case of serious burns, and artificial heart valves can be made from pigs' heart valves. Large amounts of enzymes for all sorts of different purposes are isolated from the organs of domestic animals.

Producing raw materials for a large variety of products is another essential aspect of the use of domestic animals. Leather from the skins of slaughter animals has always been used for a wide range of purposes other than clothing. Today these extend from covering upholstered furniture and car seats to the basic material for bags, suitcases and bookbindings. Goat leather was at one time converted to parchment, an important writing material. The whole skins of the small ruminants, goat and sheep, were used as containers for water or wine. The horn of horned animals can be turned into buttons, combs and ornaments, and in earlier times it was also part of the manufacture of musical instruments and tools. Hair, in addition to the central role of wool, was and is used in musical instruments (e.g. horsehair for covering violin bows), to make paintbrushes and brushes, to manufacture felt, to twine cords and finally as an insulating stuffing for cushions or mattresses. Poultry feathers also serve similar purposes and one should remember the former use of goose quills as writing utensils and ornaments.

The blood of slaughter animals, formerly a food, is now also a source for albumen adhesives, blood meal for animal feed, and pharmaceuticals. Animal fats not used for food may be materials for cosmetics, glycerine and fatty acids. The last of these are used in producing stearin candles, while glycerine is used for technical purposes, e.g. anti-freeze and pharmaceuticals, and is also the basis for dynamite production. The bones left after the removal of skin and meat form the basis of the manufacture of glue, gelatine and bonemeal for livestock feed and fertilizer. Even the inner hollow organs do not go to waste: the gut, stomach and bladder can be used as skins in sausage-making, gut for stringing musical instruments, and catgut (nowadays sheep gut) for stitching after operations as it can be absorbed by the body. Sinews have also been used as bowstrings or thread, fat as a lubricant. Finally the dung of domestic animals, apart from being dried to use as fuel as already mentioned, is a natural fertilizer.

The amount of work done by domestic animals is of inestimable significance. Large ungulates provided power for transport for thousands of years, which was a decisive factor in war and peace (Fig. 1.9). Horsemen made history in extended campaigns of conquest. Besides the horse, the dromedary attained importance as a saddle animal, particularly amongst the Arabs. Transporting goods over terrains impassable to waggons or sleds could only be accomplished with the aid of pack animals before modern technical means had been developed. Here one need think only of the donkey, and the mule (the cross between donkey and horse), in Europe and the Near East. Any significant kind of trade between the East Asian world and Europe would have been impossible formerly without the large camel caravans through the deserts of Central Asia. The yak made transport possible in the high mountains of Inner Asia as did the llama in the Andes of South America. Cattle were of central significance as draught animals for thousands of years. A reminder of this is conserved for tourists in Europe on the island of Madeira, where oxen-drawn sleds still frequent the plastered streets of Funchal (Fig. 1.10). Zebus are used as draught animals more frequently in tropical lands. The horse was employed to draw war chariots before it came to replace cattle more and more for everyday and agricultural

Fig. 1.9. Besides the part they play in satisfying the basic human need for food, domestic animals have been kept to perform work since ancient times. The horse attained special significance in war and peace in early historical times. This is expressed in the wide variety of depictions in the art of many peoples (relief from Classical Greece: National Museum of Archaeology, Athens).

purposes. Draught animals were necessary in agriculture not only to draw waggons and carts but also for ploughing fields. Donkeys and camels plodded constantly around deep wells to turn winches to draw water for drinking or to irrigate the fields. In a similar way, large ungulates such as oxen were used to help in threshing by treading out the grain from cereal plants spread out on the ground. Reindeer were used to draw sleds and for riding in the far north of Eurasia. Finally, human penetration so far into the icy wastes of the Arctic and Antarctic would have been quite unthinkable without sled dogs.

Some kinds of domestic animal, in particular the dog, perform some tasks by using specific modes of behaviour rather than physical strength. Geese can be used as guards; so too can dogs, which in a number of countries are indispensable helpers in keeping large herds of ungulates together. In hunting, different kinds of specially bred hound chase and catch the game, drive and hold it at bay until the hunter arrives, set and so locate it, seize it, fetch the bag, look for and follow tracks, as required (Fig. 1.11). Especially powerful types of dog were also bred for comba- tive purposes, above all to fight large beasts of prey. The fighting and tracking abilities of dogs in the hunt led to similar employment elsewhere. Police and Customs and Excise dogs carry out important tasks particu- larly nowadays because the efficiency of their sense of smell allows them to be trained to detect hidden goods, such as smuggled drugs. Dogs have also attained importance in searching for people buried by avalanches. Pigs can be trained to a certain extent to search for objects by smell, and, in addition to individual cases of hunting, they may aid in the search for truffles. Finally, in communications, carrier pigeons have attained a special value.

Fig. 1.10. The oxen sleds preserved for the tourists in Funchal on Madeira amidst the cars of the modern world are a reminder of the time when cattle were the most important draught animals in the daily life of the peasants of Europe as they still are today in many parts of the tropics.

Another important job performed by domestic animals is the control of commensal rodents (i.e. those living in and around houses and settlements), which cause considerable damage in stores of foodstuffs. The domestication of two kinds of animal, the cat and the ferret, was probably originally connected with this problem.

Medical research is another field no longer conceivable without the use of domestic animals. Huge numbers of laboratory animals, typically rabbits, rats, mice, hamsters and guinea pigs, are used (Fig. 1.12). Large numbers of dogs and cats also serve in testing new operating techniques and new medications, or for the general promotion of research into new problems. Breeding special miniature pigs has made it possible to use this species, normally too difficult to manage, in laboratories also.

Today more than ever, domestic animals have social functions to fulfil. Bloodthirsty forms of entertainment such as bullfighting and cockfighting are still significant, if only in geographically limited areas, whilst dog-fighting and badger-baiting have become commoner in recent years.

Fig. 1.11. Hounds were bred to help men in hunting as far back as antiquity. Large breeds were used to bring strong game to bay (Roman relief: Roman-Germanic Museum, Cologne).

Fig. 1.12. Small domestic animals are kept in huge numbers in batteries of standardized cages as experimental animals for medical research.

Greyhound racing is a form of sport involving the performance of the animal alone, while horse racing combines the active performance of both man and animal. Riding has gained greatly in significance as a recreation especially very recently: the horse is not only a 'sporting appliance' but also a companion.

The role of companion, significant mainly for children and people living alone, is also fulfilled by countless pet dogs and cats, rabbits, guinea pigs and golden hamsters, depending on how large is the available space. The wide distribution and statistical significance of such social uses of domestic animals represents an economic factor of considerable importance. Breeding horses, dogs, cats and other small animals as pets comes under this heading as do the local pet shop, the pet food industry and the manufacture of accessories for riding or for the care of dogs and cats. It should be mentioned in passing that the increase in such social animal-keeping in crowded urban areas has also created new communal problems, such as the need to provide riding paths and the removal of dog droppings from streets, public parks and children's playgrounds.

Another frequent social function of dogs is without doubt that of a status symbol. One example is sufficient: the 'status dog', whose existence (and the fact that it is taken everywhere possible) is intended to increase the esteem or at least the self-esteem of its owner. Like all status symbols it is subject to changes in fashion: it becomes more valuable if it draws attention by its conspicuous size or coat (e.g. the Afghan hound) – or, conversely, by its conspicuous smallness (e.g. the Yorkshire terrier) – and then loses its value when the fashion has become so widespread that its main purpose, attracting attention, is no longer achieved. In the more practical world of peasants and above all herdsmen, it is not the single domestic animal which has to attract notice, but rather the size of the herd, which indicates the wealth of the owner directly. This aspect has been represented by the significance of cattle, sheep, goats or camels as direct means of payment, for instance as a bride-price, for thousands of years. Large ungulates serve as signs of wealth not only for the living but also even the dead. Even today, amongst the herdsmen of southwestern

Fig. 1.13. Domestic animals have been companions for humans since early times. This relief from Classical Greece depicting a dog and a cat in a domestic scene is one of the first records of the existence of domestic cats in Europe (National Museum of Archaeology, Athens).

Madagascar, large walled tombs are painted with scenes from the life of the person buried within, depicting his wealth in zebus, and zebu horns serving the same function are laid on the roofing (Fig. 1.14).

This leads us to a last central use of domestic animals, especially in earlier civilizations: their function as cult objects. Animals kept mainly for their meat probably played an important role as sacrifices in religious ceremonies right from the beginning of their domestication. There are manifold indications of this in the surviving records of the advanced civilizations of the Mediterranean and Near Eastern regions. Domestic animals themselves also became objects of religious worship, sacred 'untouchable' beings that were revered within the cult. According to a particular characteristic, such as special strength or marked defenceless-ness, domestic animals were taken up into the language of symbols. The bull, for instance, plays a central role in many eastern Mediterranean civilizations – the Apis bull as the manifestation of the Egyptian sun god, the Minotaur from Crete in Greek mythology, with an obviously central significance in Minoan cults, the 'golden calf' in the religious history of Israel. The lamb, on the other hand, became a symbol of Christianity. Finally, domestic animals found their way into the various forms of late belief in magic, for instance the black cat as the witch's companion in the conception of the world current in the late Middle Ages in Europe, the practice of immuring animals in building foundations and so on.

Fig. 1.14. Domestic animals as signs of wealth in pastoral civilizations can contribute to the renown of their owner after his death. Tombs in the steppes of southwestern Madagascar are decorated with illustrations from the life of the deceased and indicate his possession of zebus (at the right). There are zebu horns lying on the loose rock roofing of the tomb.

Synopsis

Domestic animals are kept and bred for constant use in and around the home. They satisfy human basic needs for food, clothing and warmth, supply raw materials for many other different products, help in

performing work, are used in research for medical progress and occupy an important place in human social life. The course of history has been inseparably interwoven with the possession of domestic animals since the Neolithic, at least.

2 | Diversity of appearance

Compared with the corresponding wild species, domestic animals as a rule display a much larger diversity of characteristics. Particularly striking at first glance is the great variety of their coat colours, wild species mostly having much more uniform coats. Throwing more light on this colour change may provide insight into how this diversity came into being in domestic animals and into the context in which it took place.

Like all other characteristics of living beings, coat colours are due to a number of different genes, the bearers of hereditary information, which control the periodicity of the pigment deposits in the hairs, the degree of dominance of the two pigment types which produce blackish-brown (eumelanins) or yellowish-red (phaeomelanins) shades and the distribution of pigment at the various locations on the coat. Each of these genes can, in turn, occur with different information in the form of different alleles. Such alleles are variations in the gene's basic information that is directed towards a specific goal in the structure of the organism, e.g. the basic information on colour, which can appear as red, yellow, green or blue. In this sense, there are more or less numerous variations (mutations) of the individual genes that contribute to the colouring of a mammal's coat.

The *agouti* locus causes a banding of each single hair by spreading the pigment in periodic zones of phaeomelanins and eumelanins instead of evenly over the whole length of the hair (Fig. 2.1). Depending on the number and width of the bright and dark bands, the overall appearance of the coat is greyish-yellow to brown, while the effect, when examined at close quarters, is irregular because of the close proximity of extremely small yellowish and blackish spots or stripes of colour. This locus, named after a rabbit-like South American rodent with a coat coloured in this way, also regulates the pigment distribution in the various areas of the body as, for instance, the bright belly colour next to the darker colour of the flanks and back (Fig. 2.2). A variation of this gene in the form of a *non-agouti* allele is known for a large number of mammals. It has the effect of eliminating the zonation of pigment distribution. Yellow and black pigments are spread together over the whole hair, with the concentration decreasing from the tip to the base (Fig. 2.1). Moreover, there is no colour differentiation in different areas of the coat. This gives the impression of a fairly even black over the whole coat (Fig. 2.3) which lightens to greyish-brown only where less pigment per hair is formed as a whole, i.e. where the *agouti* colouring looks more or less white, such as on the belly of many species. Most pure black variations of many mammalian species, e.g. the normal black cat, result from this allele of the *agouti* locus. The *non-agouti* allele is recessive with respect to the *agouti*

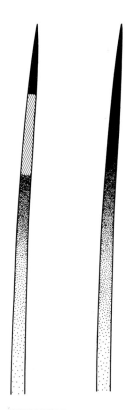

Fig. 2.1. The *agouti* locus of coat colour causes banding of the hairs (left). A dark tip is followed by a more or less broad yellowish zone and then a blackish-brown colouring which thins out towards the base of the hair. In some species, this banding is repeated several times. The mutation of this gene into the *non-agouti* allele entails a uniform deposit of pigment in the hair making the whole hair look brownish-black with the colour concentration decreasing towards the base (right).

Fig. 2.2. The greyish-yellow to brownish overall impression of the coat of an animal with the *agouti* gene arises from the close proximity of extremely small yellowish and blackish spots or stripes of colour such as result from the incomplete superposition of banded hairs standing one behind the other (domestic rabbit).

type, which is the colouring of the wild species. So, if an animal inherits the *agouti* allele from one parent and the *non-agouti* allele from the other (heterozygous state) it will have wild-type colouring. Only if it inherits the *non-agouti* allele from both parents (homozygous state) will its colouring be black.

The *albino* locus determines whether any pigment is formed at all and, if so, how much. The wild-type allele of this gene allows free play to pigment production, while the *albino* allele prevents it entirely, in which case the coat looks pure white (Figs 2.4 and 2.5). Since not only the hair but also the skin has no pigment whatsoever, the eyes of such albinos look red because the blood vessels shimmer through or bluish if the cornea is very thick. The *albino* allele also is recessive and so must be present in double dose to have any effect. There are a number of other alleles intermediate in their effects between the *albino* and the wild-type allele of the relevant gene, which decrease the pigmentation in varying degrees without suppressing it completely. Of these, it is worth mentioning the *chinchilla* allele, which is called after the rodent from the Andes which is important in the fur trade and is so coloured. The light-grey shade is evoked by a greater reduction in the yellowish phaeomelanins than in the blackish eumelanins. The pigment suppression in chinchilla-coloured domestic cats is so strong that there is only a blackish-grey gleam on the back while the rest of the coat is whitish. The *chinchilla* allele in rabbits, on the other hand, results in an ash-grey coat more like the species from which it is named. In all, six alleles at the *albino* locus are known in rabbit breeding. As is the general practice in genetics, these are symbolized by letters. The letter '*A*' designates the wild-type allele and is capitalized to express its dominance over the recessive *albino* allele '*a*'. The *chinchilla*

Fig. 2.3. The coat of an animal with the *non-agouti* allele looks black or blackish-brown (domestic rabbit).

allele is termed a^{chi}. There is, therefore, the following allele series of progressively reduced pigmentation in the rabbit: A (full pigmentation); a^d (dark chinchilla); a^{chi} (chinchilla); a^m (light chinchilla); a^n (Himalayan); a (albino). The Himalayan colour preceding the albino and corresponding to the Siamese colouring in the domestic cat involves a high sensitivity to temperature in the colouring of the coat. Pigmentation is lacking to a large extent in coat areas close to the centre of the body, where the skin temperature is constantly higher. The young, coming from the uniform warmth of the womb, are born completely white. The parts where the skin temperature is predominantly lower, i.e. the snout, the ears, the legs and the tail, become darker as the animals grow so that a

Fig. 2.4 (above) and Fig. 2.5 (below). The normal, yellowish-grey colouring of the wolf (above) has disappeared in primitive dogs in favour of reddish to yellowish-brown shades determined by phaeomelanin (here Australian dingoes, below). Black, white and white spotting, however, also occur. The white dingo in the lower plate is an albino without any pigment in its coat. Since this is also the case in the irises of its eyes, the incidence of light in its eyes is too strong for the animal when it is very bright and so it has to squint.

dark mask forms on the head, as is characteristic for the Siamese cat. The original white of the young Siamese cat's body also turns into a cream to light-brownish colour due to subsequent weak pigmentation as it grows. If such an animal loses hair on any particular part of its body and it grows again when the external temperature is cooler and thus the affected skin area is also cooler, this hair is darker like that at the extremities of the body. So a Siamese cat that spends much time out of doors in winter is darker then than one kept constantly in heated rooms.

In addition to the *agouti* locus, the *dilution* locus regulates the distribution of pigment in the hair. The wild-type allele elicits the normal, uniform pigment distribution; the *dilution* allele leads to pigment clumps which leave other sections free of pigment so that the overall coat colour becomes lighter. A black based on the *non-agouti* allele of the *agouti* locus combined with this *dilution* allele, for example, turns into a pale bluish-grey shade usually termed 'blue'.

A *black* (*intensity*) locus controls the formation of eumelanin. The allele which reduces the ratio of black pigmentation, combined with the *agouti* allele of the *agouti* locus, leads to a more yellowish or reddish, sandy colouring, depending on whether the phaeomelanin is more yellow or more red. Combined with the *non-agouti* allele, the black of the coat lightens to brown. An allele at another locus determining the spread of the black pigments causes the yellowish-red phaeomelanins alone to determine the animal's appearance (Figs. 2.4 and 2.5). The opposite is also possible, i.e. strengthening of the eumelanin production as compared to that of the normal wild colour. Even combined with the *agouti* allele this results in blackish colouring of the bands on the hair, producing a shade known as 'steel-grey' in rabbit breeding.

As the *black* genes alter the eumelanins, so the phaeomelanins are changed by the *yellow* (*intensity*) genes, which are decisive in the toning of the yellow and red shades. *Silver* genes, when the corresponding *silver* alleles are present, are responsible for the occurrence of hairs without any pigment, i.e. white hairs, amongst normally coloured ones. The intensity of the silver gleam throughout the coat depends on the number of white hairs. This number increases in the course of juvenile development. Silvering with age is also known in connection with the *grey* genes in horses. White horses are born dark and then grow more and more white hairs with each change of coat over the years until the completely white coat is fully developed. Going grey with age in humans is a similar process.

The *white-spotting* locus is one of the few colour factors where the colouring deviating from the wild type is as a rule dominant, at least partially, insofar as it shows in the form of some white spots when even only one of the relevant alleles has been inherited. However, one may speak of intermediate heredity here, as the white portion of the patched coat in the homozygous state (i.e. when the *white-spotting* allele alone is present in double dose) is much more extensive than in the heterozygous state when the allele for normal colour is present in addition to the *white-spotting* allele. Modifying genes occurring in addition to the main gene of the *white-spotting* locus determine the degree of large-sized spotting. In the extreme case, there are only a few normally pigmented coat areas (Figs. 2.6–2.8).

An allele series of the *pattern* locus comes into play in species with a pattern in addition to the basic coat colour. The domestic cat serves as an example of this. European domestic cats usually have a coat pattern of stripes running crosswise to the longitudinal axis of the body which are interrupted to various degrees, with the stripes ultimately becoming rows of single spots (Fig. 3.13, p. 47). In addition to this striped tabby pattern, there are domestic cats with a spotted pattern in which the flanks are completely covered with separate spots, characteristic of the race called Egyptian mau. One very frequent variation of the tabby pattern is the so-called blotched tabby, in which an increase in the number of coat areas with stronger pattern pigmentation completely reverses the pattern type (Fig. 9.8, p. 138). Finally, there is a fourth variant in domestic cat patterns, the Abyssinian tabby type characteristic of the so-called Abyssinian cat. Here the pigmentation of the pattern is suppressed to the point where slight remains of markings still appear only on the head, limbs, tail and lower flanks.

The extreme diversity observable in the coat colouring of domestic animals depends not only on the co-existence of many such alleles of the colour loci, but also on the numerous possibilities resulting from their combined occurrence (Fig. 2.7). In the cat, the *white-spotting* locus with its corresponding allele causes white areas on the otherwise wild-

Fig. 2.6 (left) and Fig. 2.7 (right). White spotting covers varying proportions of the coat, sharply set off from the normally pigmented areas. In combination with other coat colours differing from the wild colouring, white spotting can lead to very colourful appearances, as in this cat and dog. The colouring of this 'three-coloured' cat is due to an *orange* allele of a sex-linked colour gene in addition to the *white spotting* allele. This gene has its locus on the X chromosome which occurs doubled (XX) in the female but only singly (XY) in the male. So only female cats can simultaneously bear the alleles for orange and for normal colour, which are then expressed side by side in a kind of spotting.

coloured, brownish-grey coat with a striped pattern. If the *non-agouti* allele at the *agouti* locus is homozygous, i.e. contributed by both parents, the result is an animal spotted in pure black and white. Further combination with the eumelanin-reducing allele of the *black* locus leads to red and white spotting. Finally, adding the *dilution* allele results in a cream and white cat. Anyone who keeps domestic cats is able to achieve a wide-ranging palette of the most varied colourings and shadings by planned crossbreeding. Such possibilities have been widely exploited in the breeding of coloured minks for fur fashions (Fig. 2.9).

The frequency of colouring deviating from the norm in wild animals forms a marked contrast. The differences between individual animals within a regional population are mostly only slight. Immediately noticeable differences, such as albinos, do occur repeatedly in most species but

Fig. 2.8. In domestic animal breeds not subject to improved breeding or colour selection, there is usually a diversity of colours within a single herd. This is shown here for zebus, the slender domestic cattle of hot dry regions, which are better able than other groups of breeds to bear the heat (steppe in southwestern Madagascar).

Fig. 2.9. A large palette of the most varying shades of colour in ranch mink displays the diversity attainable by selective breeding to combine a large number of alleles of the various coat colour loci.

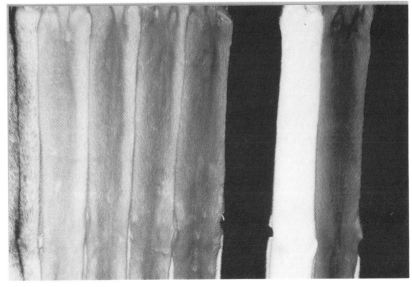

remain extremely rare on the whole. The only limited exception is black colouring. For instance, there are populations of squirrels consisting exclusively of red to brown animals in Central Europe, others with sporadic black specimens and finally ones in which black is the normal coat colouring determining the appearance of the whole population. Another well-known case is the black variant of the leopard, frequently called a panther, which occurs with increased frequency among its normally coloured conspecifics that have coats of black spots on a yellowish background, especially in regions of Southeast Asia. Examples of such frequent occurrences of black colouring (melanism) can be cited for most groups of mammals.

A wild animal's colouring is usually well adapted to its natural environment. For instance, the majority of African wild cats from the hot, dusty, desert steppes of North Africa and the Arabian Peninsula have pale sandy or pale brownish coats on which the patterning often appears only indistinctly or has practically disappeared completely due to the *Abyssinian* allele described above for the domestic cat. These wild cats then appear pale in their surroundings. In the more humid savannah lands further to the south of Africa, the dark brownish-grey African wild cat with a distinct pattern is predominant (Fig. 3.12, p. 46).

Colouring matching the background by prey animals makes it more difficult for predators that orient themselves mainly by sight. Such camouflaging is particularly striking in animals that look as though they would be impossible to miss once they have been removed from their natural surroundings. An example is the ring-tailed lemur from Madagascar, with markings particularly rich in contrasts. Its black and white banded tail and black and white face mask against the greyish-brown of the rest of the body give it an appearance which strikes the eye immediately in a zoo. In its natural habitat, the sparse, dry forests of southern Madagascar, the sun shines almost constantly through the leaves and branches of the trees, producing such a sharply contrasting pattern of spots and stripes of glittering light and deep shadow on the ground that the ring-tailed lemur seems to dissolve into its surroundings (Fig. 2.10).

Such colouring and patterning, which optically resolves the animal's body and camouflages it in its environment, are the result of corresponding selection that provides the animal best adapted to the particular ecosystem with the best chances of survival and reproduction. Individuals with conspicuous colouring that occur now and then in a population are more readily detected by hunting predators and so are in greater danger. Animals deviating from the prevailing norm are threatened by a further selection process in the case of species living socially. It is difficult for a predator to concentrate on a single victim when a whole band of animals, all of which look very much the same, scatters before its eyes. If, however, one amongst them looks distinctly different, it immediately becomes the preferred target as it is easy to follow. In this way goshawks pick individuals differing from the mass out of flocks of feral pigeons; that is to say white ones when the majority is dark or vice versa. There is a third source of danger for oddly coloured animals living in social groups, namely social isolation and aggression from conspecifics. Just as the odd or conspicuous specimen is an outsider for the predator, it may also become one right from the start for the rest of its group and be forced into a low rank order with only slight chances of reproduction. As long as its being different is not noticeable nothing much happens to it in this respect. This was probably so for an albino freshwater dolphin female while it grew up in the murky, muddy waters of the Orinoco. After it was caught by an expedition of the Duisburg Zoo along with several normal (darkly coloured) conspecifics, and kept with them in clear water in the zoo, it gradually became the object of more and more social aggression from other dolphins, which finally resulted in its death (Fig. 2.11). The combined effect of all these forms of selection – selection of the conspicuous by predators, selection of the deviant from the norm in a social group by predators, and social aggression on the part of conspecifics – leads to a certain standardization of colouring being constantly maintained amongst wild animals.

In domestic animals, the first two of these important selection factors are eliminated as defence from predators is taken over by humans. Decreasing selection immediately allows diversity to increase. Where predator pressure has been reduced because hunting by humans has exterminated or greatly depleted carnivores, the result is the same in

Fig. 2.10. Colouring serves the wild animal mainly as camouflage in its environment. This is also the case for animals that look extremely conspicuous in the altered environmental conditions of zoos. The contrasting black and white head mask and black and white banding on the tail camouflage the Madagascan ring-tailed lemur particularly well in the sharply contrasting play of light and shade on the ground and between the leaves in the sparse dry forests of its distribution area.

principle. For example, since the great decline in the numbers of birds of prey in the cities of Central Europe, there has been a noticeable increase in the occurrence of pied and albino blackbirds and sparrows. White and strikingly light coloured fallow deer and pied roe deer have likewise been able to hold their own in Central European hunting grounds since the wolf and lynx have been exterminated.

The basis for the limited variation in wild animals and the diversity in domestic animals is the same. Mutations enrich genetic diversity. Selection in the opposite direction constantly maintains the standardization in wild animals to a large extent. This selection is reduced in domestic animals, so the standardization constantly decreases. If selective breeding is added to the resulting enriched diversity, new, altered standards become possible. Of course, this mechanism is valid not only for colouring but for all complexes of characteristics. So not only variation in colour but also in the shape and size of the hairs of the coat increases strikingly.

Fig. 2.11. Animals of a colour strikingly different from that of their conspecifics are in danger of being isolated as social outsiders and becoming the objects of aggression. This albino freshwater dolphin from the Orinoco eventually died because of the aggression it suffered from its darkly coloured conspecifics. Its striking colour was clearly visible in the filtered water of the zoo aquarium (Duisburg Zoo), whereas it would not have been nearly so obvious in the murky waters of its original habitat.

Both long- and short-haired types of most domestic animals are known. Long-haired types are often called angoras, e.g. angora cats, angora rabbits, angora guinea pigs (Figs. 2.12 and 2.14). This name comes from the long-haired Turkish angora goat known from the area around Angora – the modern Ankara – from which mohair is obtained. As a rule, long hair is passed on recessively, so that mating a long-haired with a short-haired animal usually produces short-haired young in the first generation. Another occurrence of long-haired animals is not to be expected until the second generation, when they are homozygous if mated together. Another deviation from the normal hair type is crinkled or tightly curled hair, as is familiar in the poodle and the karakul. The latter provide the broadtail or Persian lamb important chiefly for the fur industry. The composition of the coat of domestic animals often differs from the norm for mammals, which consists of long, straight guard hair and a woolly

undercoat. Reduction of the underfur, e.g. in the Maine coon, Norwegian forest and Turkish van cats, as of the guard hair (e.g. in rex cats), occurs in domestic cats. Reduction of the guard hair and increase in the woolly undercoat is the basis for the most important source of wool, the wool sheep, which are to be distinguished from the so-called hair sheep. Finally a total or far-reaching reduction of hair, leaving residues only on isolated body areas, leads to naked animals such as hairless dogs and cats.

The skin itself, where the hair originates, also proves to be very diverse in domestic animals. Sometimes the skin increases, leading to the formation of creases, as in the faces of some breeds of pig (Fig. 2.17) or of a huge dewlap, as in the zebu. Also, lop ears occur in many domestic animals, the best known being those of the dog and rabbit.

Differences in size are especially striking in domestic animals. In some

Fig. 2.12. The name angora for long-haired domestic animals comes from the angora goat of the Ankara region.

Figs. 2.13 and 2.14. Short-haired and long-haired (angora) guinea pig demonstrate how the type of hair alone can change the appearance of an animal species totally.

species, they far exceed what can be observed in their wild ancestors, the scale ranging from extreme giants to extreme dwarfs. Shoulder height in dogs can vary up to fourfold, and in horses nearly as much. Weights differ to an even greater extent. Such variations in size are inevitably accompanied by certain changes in proportions. Regular shifts, in which the size of a part of the body does not alter in a constant ratio to the whole body size, i.e. is not isometric, are called allometric. Positive allometry means that a part of the body becomes both relatively and absolutely larger with respect to an increase in overall size, negative allometry that the body part increases in absolute size but decreases in relative size. The muzzle is usually affected by a very distinct positive allometry. The larger

Figs. 2.15 and 2.16. Curling of the hair is characteristic of the karakul breed. This curling and waving is not developed in the coat of prematurely born lambs, which have a flat moiré pattern instead and are known as broadtails in the fur trade (Fig. 2.15, upper figure). The curls are developed in lambs a few days old but are still firm and closed. They provide the so-called Persian lamb (Fig. 2.16, lower figure).

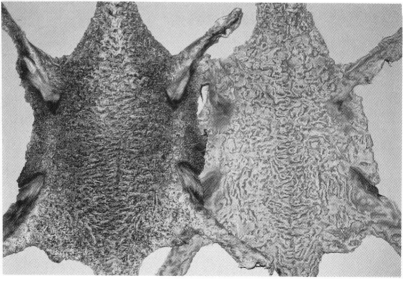

the animal the longer is its muzzle compared to the braincase, so that it dominates the head to an increasing extent; the smaller the animal, the more the muzzle decreases, leaving the braincase to dominate (Fig. 2.19).

This marked rounding of the head in dwarf forms and correspondingly in the young of larger races, leads, by combining a high forehead and large eyes, to the 'cute' response – an intensification of care behaviour in humans, particularly women. This is what Konrad Lorenz called the

Fig. 2.17. Skin folds characterize the very shortened face of some pig breeds from Southeast Asia, like the small Vietnamese domestic pig pictured here, which has become popular under the name pot-bellied pig in European zoos.

Fig. 2.18. Enlarged areas of skin are found in many domestic animals in the form of lop ears (Saint Bernard).

Fig. 2.19. As a consequence of regular shifts in the body shape as the body size decreases (allometric changes), the muzzle is greatly reduced in the overall appearance of the face in miniature dogs while the forehead becomes higher and rounder and the eyes appear relatively large. Such animals fit the pattern of baby-like characteristics as described by Konrad Lorenz, which appeals particularly to women and intensifies care behaviour. The breed depicted here (Yorkshire terrier) is distinguished from dogs of normal growth by another proportional alteration independent of size, namely its short legs.

Fig. 2.20. The sloughi, a Moroccan greyhound, embodies a type of physical structure different in every respect from the Yorkshire terrier shown in Fig. 2.19. It is distinguished by a particularly slender growth and very long legs.

infantile schema. It is probably due to this behavioural element, of which the person in question is completely unaware, that some elderly women choose a cat (which generally has a higher forehead and shorter muzzle than a dog) or a small or miniature dog as a pet.

In addition to such changes in proportions due to size, there are others unrelated to size which are even more striking. The whole growth type of the body may alter, causing slender or broad types of growth such as characterize the relationship between thoroughbred and heavy-breed horses. Greyhounds (Fig. 2.20) are the corresponding slender types in the dog, and the zebu (Fig. 2.8) amongst cattle. There are short-muzzled forms of most domestic animals, independent of the allometric change in muzzle length due to size. Such forms are sometimes called pug-headed, which alludes to the breed of dogs with this type of head. Another typical short-snouted breed is the boxer. Amongst the pigs, the extremes are mainly to be found in races of East Asian origin such as Chinese pigs and the Vietnamese pot-bellied pig (Fig. 2.17). The breeding of Persian cats runs in this direction, and the Arab horse has a certain tendency to being short muzzled. Changes in leg proportions result in short- and long-legged types. In addition to numerous short-legged breeds of dog, of which the dachshund is typical, there are also short-legged cattle and sheep. The opposite, long leggedness, is characteristic of, for example, greyhounds and thoroughbred horses. Alterations in the vertebral column frequently occur in the tail area, where short and curly tails and taillessness are widespread. Increasing the number of vertebrae in the loin region has been practised in pig breeding to increase the number of loin chops.

As with the body shape, there is considerable diversity in the horns of domestic horned animals. Two fundamentally different horn types can be distinguished in European cattle as far back as the Neolithic: the so-called 'brachyceros' (or shorthorn) and the 'primigenius' type with longer horns (from the species *Bos primigenius*, the aurochs or urus (Figs. 2.21 and 2.22)). From these two, several different forms have developed as a result of further genetic changes. The extremes are represented by the giant horns of some East African cattle and by hornlessness. Hornlessness is also one of the possibilities in sheep and goats. In addition to the flat spiral type, which is considered to be the original shape, sheep may have straight corkscrew-shaped horns, e.g. the Hungarian zackel sheep (Fig. 2.23), and even four horns (Fig. 2.24). There are two basic types of goat

Fig. 2.21. Some domestic ruminants have horns of very diverse sizes and shapes. Hungarian steppe cattle, for example, have very long horns.

Fig. 2.22. The Central European cattle breeds are representatives of the short-horned cattle, such as the Alpine cow shown here (brown Swiss).

horns: sabre shaped, such as is found in the wild bezoar goat, and corkscrew type, with very broad spiral whorls (Figs. 3.39, (p. 72), 3.43 (p. 74)).

This colourful and wide-ranging diversity in the overall appearance of domestic animals has provided the basis for the development of a multiplicity of breeds, of which those that are highly improved are easily

Fig. 2.23. In contrast to the original curled type of sheep horns, the Hungarian Zackel sheep has straight horns with corkscrew twists.

Fig. 2.24. The horns of the four-horned Jacob sheep, an old English breed, are split at the root and grow as double horns giving the impression of four in all.

recognizable. They are groups of individuals of a species that resemble each other strongly in a series of characteristics regarded as typical for a breed but are distinctly different from other individuals of the same species with respect to the special combination of their characteristics. This restriction of the original diversity of domestic animals in the direction of a new stabilized uniformity, a 'typical' expression of a set of characteristics, is achieved by selective breeding according to a standard agreed upon by the breeders. The standard itself can be based on ideas of beauty or arise as a result of certain practical needs. Since both human ideas and needs are subject to change, far reaching alterations in any one breed of domestic animal can be observed even within the course of a century as a result of the breeding purposes having altered.

A case in point is that of breeding Persian cats. Their origin lies in the angora cats from the Asia Minor–Near Eastern region; they had a normal body shape but longer hair with a certain tendency to a ruff and a bushy tail. In the course of the present century, the breed now called Persian cats, with a thickset growth form and sturdy legs, a somewhat shorter snout and an arched, broadened forehead as well as small ears, was developed from the angora cats. At the same time the length of the hair was increased still further. This breed having departed so far from its original of past decades, some breeders have now acquired unaltered angora cats in Asia Minor and developed a new standard for them. This illustrates the splitting of one race into two, one of which was bred by changing the breeding goals for several features, while the other preserves the original appearance of the race more or less unchanged.

To change the appearance of a breed, not only is the practice of strict selection in accordance with new breeding goals useful but there is also the possibility of attaining what is desired by crossbreeding with one or more other breeds, a method which plays a decisive role in modern animal breeding. In the dog, its effect can be followed in the history of the change from the original Saint Bernard to the modern breed. The original type, the hospice dog of the early nineteenth century, was a large, thick-haired Swiss alpine shepherd dog of normal build. New breeding goals turned it into the still larger, long-haired, modern Saint Bernard with an upturned muzzle and consequent snub-nosed profile, by interbreeding with the Newfoundland. In pig breeding, the local breeds from Central and North Europe, originally bred for bacon, became distinctly meat pigs with longer bodies, through breeding and selection to meet the altered requirements of the meat market. After 1850, the original landrace pigs were crossbred with breeds imported from England that, in turn, had been crossbred with pigs from East Asia since the end of the eighteenth century to produce the so-called improved landrace pigs and large whites.

So far, the discussion has concerned breeds with a narrowly defined appearance selectively bred to achieve a certain standard. Where this is not the case, as in primitive local breeds and in the diverse 'mongrels' at the primitive level of domestication, it is extremely difficult to make a distinction. The problem to be resolved is similar to that involved in attempting to subdivide a wild species, with the intraspecific diversity necessitated by the dynamics of evolution, into the static categories of races or subspecies. The only way to describe a subspecies of a wild animal species or a breed of a primitive domestic animal not bred by man to produce a certain standard type seems to be in terms of the most typical characteristics. In such a case, a single fixed type or standard alone is insufficient; the definition has to be constructed either on the frequency of characteristics or on the frequency of the underlying respective alleles in the gene pool of a whole population. This is the sense in which the concepts of primitive and improved breeds of domestic animal can be contrasted, though it must clearly be kept in mind that in reality the two constantly overlap.

Just as in the origin of first differences in geographically separate populations of a wild species, the first shifts in gene frequencies in a domestic animal occur in the phase of its spread – in the wild animal when it actively colonizes new areas, in the domestic animal when it is spread by human agency. In addition to the effects of selection, genetic drift and crossbreeding with related wild species are responsible for the initiation of new breeds that arise during this process. Genetic drift is the random shift in frequencies of alleles and in characteristics in the course of the foundation of new populations from the spread of only a very small proportion of the original population. The few founders bring only a random part of the whole genetic diversity of the original stock along with them. If such a process of spread takes place repeatedly it is extremely unlikely that the genetic drift will be in the same direction in all cases. If only a few animals of a species are transported to new places, one or other expression of characteristics from the original diversity of the large initial population will predominate in each case. The propagation of such very small groups consisting of only a few individuals to form first local and then, after further increase and spreading, regional populations leads to genetic units differing in individual traits that may be designated as primitive breeds. A new influx from the initial population or a meeting and mixing with a small founder population from a neighbouring area will again wipe out such differences.

Another way that the different primitive races of a species of domestic animal can arise is via crossbreeding with wild conspecifics. As some examples in the next chapter will make clear, the ancestral species usually consist of several geographically separate populations differing from each other in the markedness and frequency of some or many features, and commonly classified as so-called subspecies in zoological taxonomy. If a few individuals are removed from a single one of these geographically separate populations or subspecies for domestication, the resulting domestic animal will possess the typical characteristics of just this genetic unit. During the first stages of domestication with only a handful of founder specimens, genetic drift will lead to some random features predominating, while others will recede or disappear completely. The subsequent spread of a domestic animal that has come into being in this way may occur both in areas where the wild species lives in the same or other geographically separate populations and in areas where the wild species is absent. In the latter case, the only factors that can affect further development of populations of domestic animals from a single, small, imported founder stock are the mechanisms of genetic drift and selection. Selection can take place through deliberate beeeding to achieve certain purposes as well as naturally, as for example due to climatic factors.

When the spread takes place into areas where the wild species lives, these mechanisms, of course, come to bear but there is the further possibility of changing the gene pool by crossbreeding with the form of the wild species living in the same area. Such a process may happen by chance when a female domestic animal, wandering unattended, mates with a male of the wild species and then introduces her young into the domestic stock.

Such crossbreeding may, on the other hand, also be brought about and controlled by human agency, particularly in the catching of young of the related wild species in order to increase the stock of domestic animals more rapidly. As Sandor Bökönyi has demonstrated using extensive archaeozoological finds from excavation sites in Hungary, such a method was applied to cattle and pigs on a large scale during the Neolithic. Catching wild aurochs calves increased the originally imported stock. Then the crossbreeding of these aurochs into the domestic herds in which they matured increased the diversity of features in these cattle in the direction of the European aurochs. The same principle came into effect as wild pigs were included into stocks of domestic pigs that had originally been imported. Corresponding measures, which should be termed secondary domestication, were not possible with domesticated sheep and goats in the Hungarian region, since the relevant wild species did not occur there.

Such secondary domestication due to the additional breeding of local indigenous wild relatives of a domestic animal naturally leads to the assimilation of characteristics normally neutral in the process and state of domestication. This leads to the domesticated animal becoming more like its wild relative and contributes to the formation of different primitive breeds. If such occurrences are not taken into account in comparative studies of the subfossil remains of domestic animals from periods of secondary domestication it is easy to draw the mistaken conclusion that purely local primary domestication took place for the same species at very different times, as has in fact frequently happened in the archaeo-zoological literature.

A brief glance at the developments in the dog and the cat may serve to exemplify the combined effect of all these mechanisms of genetic drift, selection and crossbreeding in forming primitive breeds of domestic animals. As will be described in detail in the next chapter, the domestic cat is descended from the wild cat of the Egypto–Palestinian region. After primary domestication there, it was passed on into other lands, finally reaching Europe as well as East and Southeast Asia. In Europe there is a widely distributed close relative, the European wild cat (Fig. 3.10, p. 45), but there is no such directly related species in China, Indochina or Indonesia. Crosses between stray domestic cats and wild cats occur from time to time even nowadays and it is not to be supposed that this was otherwise from the beginnings of the history of the domestic cat in Europe. The original wild cat-like primitive domestic cat is characterized for example by a general darkening of the coat pattern on the back, with no sharp dorsal stripe and a low, slim, lower jaw. In comparison, the modern domestic cats in Europe and those cat populations derived from them in other countries more frequently have a dorsal stripe in the coat pattern as well as a lower jaw tending rather in the direction of the heavier form of the European wild cat. In Southeast Asia, with no closely related wild cats, selective measures led to improved breeds such as the Siamese cat, which has much more of the ancestral wild cat in its skull structure and behaviour than its European relatives. A skull type related to that of the Siamese cat exists in the cat population of Madagascar, where the frequency of individual alleles of coat colour recalls that of the cats in Indonesia, probably one of the original homelands of Madagascan civilization. The varying frequency of individual colour alleles like the Siamese colour originally from Southeast Asia or the white spotting with a high proportion of white and only slight areas of other colouring seems, however, to be a product of genetic drift and probably also of selective measures.

   As will be shown in the next chapter, the dog probably originated from the marginal zone of southern wolf populations found from Arabia to South Asia. From there, it spread to Europe, Central and North Asia, and America – lands with wolf populations differing geographically from the dog ancestor – further into Southeast Asia, Australia and the Pacific as well as to Africa, i.e. into regions without wolves. Since reddish-brown, black, white and white-spotted animals are to be found among the dingoes, and the so-called 'Hallstrom' dogs of new Guinea (Fig. 2.5), while the typical wolf-grey is entirely lacking (Fig. 2.4), and since this also seems to have been the case in the primitive dogs of Africa and Polynesia before the introduction of European breeds of dog, it can be inferred that the corresponding colour allele was lost right at the beginning of domestication. The wolf colour, however, turns up again in European breeds and is found particularly frequently in dogs from the northern countries (Fig. 2.25). It is known that wolves were quite frequently crossbred with sled dogs and resemblances to wolves are found amongst them to an increased extent in the facial features. The long hair of the northern as compared with the predominantly short hair of the southern primitive dogs can most likely be traced back to climatic selection. Differences in size, bodily structure and hair between the geographically neighbouring dingo and New Guinea dogs may originate in a combination of selection and genetic drift effects. Finally, certain differences in skull structure between local dingo populations in Australia and lines of dingoes in European zoos are probably the result of genetic drift alone. Selection by human agency with a view to use seems to be the main factor in the genesis of the poi dogs of the Polynesian islands, which are very different from the long-legged, slender dingoes and the similarly mostly long-legged New Guinea dogs of

Fig. 2.25. The wolf colouring that primitive dogs in countries without wolves apparently lacked right from the beginning turns up again in northern dog breeds from wolf-inhabited regions. It is most probably attributable to repeated crossbreeding with wolves (Greenland husky dog/Eskimo dog).

Fig. 2.26 . Two kinds of primitive dog with extremely different body shapes live in the Pacific region, the Australian dingo (at the rear) and the poi dog of the Polynesian islands bred for its meat (right foreground, after a back-crossed animal in the Honolulu Zoo). A possible transition between the two is demonstrated by the different types of bodily structure of the 'Hallstrom' dog from New Guinea (New Guinean dingo), the third primitive breed from the western Pacific region. Its normal shape (left) resembles that of a small dingo, a short-legged variant paves the way to the poi dog (centre foreground, drawn from a photo in Schultz, W. (1969), Zur Kenntnis des Hallstromhundes (*Canis hallstromi*, Troughton 1957), *Zoologischer Anzeiger* **183**, 47–72).

the western Pacific area. The poi dogs, the pure form of which has now disappeared due to mixture with European breeds but is being reconstructed in a back-crossing project in the Honolulu Zoo, were meat dogs with characteristically long, heavy bodies, short legs and very large erect ears. They were very much reminiscent of a primitive breed of dog of the Pacific coast of Mexico, and dog types in Inca pottery, thus evoking ideas of a cultural connection via the Pacific to the east. They were already on the way from a distinctly primitive dog to becoming an improved breed (Fig. 2.26).

Synopsis

In contrast to wild animals, the appearance of domestic animals is very diverse especially with respect to the colour and type of coat, and body size and shape. This diversity is based on a reduction in natural selection on the one hand and selection in the course of breeding on the other and provides the foundation for the origin of numerous breeds. Primitive breeds are formed by crossbreeding with different populations of the wild ancestral species, via chance shifts in gene frequencies and through selective influences. Improved breeds, which are more easily defined, result from strict, standardizing breeding measures.

3 | The origins of domestic animals

In seeking the origins of domestic mammals we go back in part to the Stone Age and in part to the early advanced civilizations. In the seventh millennium BC, at the latest, long after the early Stone Age when the wolf had been domesticated and had become the first domestic animal, the dog, other domestic species based mainly on the sheep and the goat and with only a few cattle and pigs came to Southeast Europe. In the course of the progressive improvement in climate up to the climatic optimum after the last Ice Age, beginning about 5500 BC and lasting until about 2500 BC, during which the summer temperatures were 2 deg.C to 3 deg.C above those of today, these domestic animals spread via the region of Hungary to Central, western and northern Europe. In the further course of the Neolithic, the secondary domestication of cattle and pigs mentioned in the previous chapter took place through increasing capture of the indigenous species.

To assess the relations of humans to animals in those periods of primary and secondary domestication, we must turn to aspects of prehistoric and early historic animal-keeping as evidenced in human activites in catching, transporting and keeping animals in the millennia before the Christian era. One of these is the part of the Gilgamesh Epic that turns up in the Old Testament as the story of Noah. The Gilgamesh Epic itself originated in the Sumerian civilization of the third millennium BC and was widespread in the civilizations between the Tigris and Euphrates. In this legend, which is basically the same everywhere regardless of the context in which it has been handed down, the hero, who survives an extensive, catastrophic flood, receives the divine command to load pairs of domestic livestock and all kinds of other animals into a large boat so as to save them from the deluge. This at first sight apparently insignificant report implies that an important principle of nature conservation was conceivable in the Sumerian civilization, namely the management of the survival of animal species by temporarily keeping them in captivity and that the idea existed of transporting domestic animals, and others, by boat. So it is not surprising that there were indeed zoo-like facilities in the Sumerian civilization.

There is even proof of such early overseas transport of wild animals in the archaeozoological situation on Mallorca. The Balearic Islands have been separated from the European mainland and the other Mediterranean islands for at least five million years. Because of the insular situation, it was possible for a unique and special fauna, a local development of Tertiary immigrants, to survive until after the Ice Age. This ancient animal world was exterminated after the appearance of humans on the archipelago in the Neolithic, and animals of a modern type were introduced. Thus, the remains of some mammalian species that developed ónly long after the last separation of the Balearic Islands from the mainland, such as hares, rabbits, cats, long-tailed fieldmice and house mice are demonstrable in the archaeological excavations on Mallorca. Fallow deer and red deer were added to these as game animals in Roman times. A number of other smaller mammals in the modern fauna of Mallorca were also most likely introduced by human agency.

A very similar situation exists in the group formed by the Lesser Sunda Islands and Melanesia. There is some evidence that there may have been pigs in New Guinea 10 000 years ago. Since this island was cut off from the mainland long before the evolution of pigs, it seems hardly likely that they reached the island as a result of natural dispersal. Along with the rusa (Timor deer), the pig has been found on the island of Timor in excavation layers dating from the third millennium BC and it is most likely that both species were introduced there.

There are also numerous pictorial representations of various kinds from early advanced civilizations in the third millennium BC, confirming that several large mammals were kept and tamed, although they did not become domesticated. This is the case for elephants in the ancient Indus civilization of northwest India; in the Sumerian civilization this holds for the catching and taming of onagers, the hemiones of the near east phylogenetically somewhere between the horse and the donkey

Fig. 3.1. The onager (*Equus hemionus*), a species to be classified as approximately between the horse and the ass, was tamed in the Sumerian civilization of the third millennium BC but apparently not domesticated.

(Fig. 3.1). The large desert antelopes (scimitar-horned oryx and addax) were kept like domestic animals in the Old Kingdom in Egypt (Fig. 3.2), where the ibex and hyena were also tamed.

All this leads to the conclusion that even in the third millennium BC, at the time of the first advanced civilizations in Egypt, in the Near East and in India, catching, keeping, taming and transporting numerous species of wild animals was widespread from the eastern Mediterranean area to the islands of Southeast Asia. It cannot be assumed that this tradition first began in all these areas just at that time; it may, at least partly, reach back still further. At least from that time on, people began experimenting with keeping many animal species including large and rather dangerous ones.

Only some of these became domestic animals. The rest apparently lacked the potential for domestication right from the start, or an initiated domestication process did not lead to any real success, or the people finally lost any further interest in the species in question for some other reason. In some cases, such as that of the cheetah kept to help in hunting, it was probably failure to breed in captivity, the first prerequisite. Alternatively it was too time consuming to domesticate an easily caught and tamed species, in preference to catching it as the need arose, as in the case of the elephant. The intended use may have involved disadvantages in connection with domestication as opposed to catching wild specimens, as was certainly the case with the cheetah and probably also with the elephant. Finally, turning a wild species into a domestic animal meant that certain pre-adaptations, certain potentialities, on the part of the animal and certain lasting interests in its use on the part of people, has to coincide. Precise knowledge of the wild ancestors of each species of domestic animal is indispensable for assessing the special suitability of the wild forms to becoming domesticated, i.e. for ascertaining the principles

Fig. 3.2. The addax (*Addax nasomaculatus*) from the deserts of North Africa is a large antelope that was kept for agricultural use in the Old Kingdom of Egypt though no domestication resulted.

of domestication. The following sections provide a brief outline of the origins of domestic mammals.

The dog

The wild species most closely related to the dog belong to the genus *Canis*, from the family of dog-like predators, the Canidae. To this genus belong the species wolf (*C. lupus*), red wolf (*C. rufus*), coyote (*C. latrans*), and the jackals (*C. aureus, C. adustus, C. mesomelas* and probably *C. simensis*). Numerous studies on single characteristics and complexes of characteristics of these species have shown that, wherever there are clear differences between the species, the dog matches the wolf better than either the coyote or the jackal. This is the case, for example, for structural features such as the form of the carnassial tooth, for behaviour such as vocalization and for biochemical features such as the serum protein pattern. So, despite some other, older views now super-seded, it is the wolf alone that comes under consideration as the ancestral form of the dog. The wolf is distributed from Europe over the Near East to South, Central, East and North Asia, and from North to Central America in a large number of very different geographically separate populations (Fig. 3.3). Evolution had progressed to varying degrees within this gigantic area. Broadly speaking, the highly developed north-ern wolves can be distinguished from the southern wolves, which have remained primitive. The latter, namely wolves from the Arabian Penin-sula and South Asia, can be regarded as surviving relic populations of an old evolutionary grade of wolf with a relatively small brain (Fig. 3.4) and

Fig. 3.3. Schematic depiction of the former distribution of the wolf. The area occupied by the northern wolves and the similar Central and East Asian steppe and mountain wolves is black. The distribution area of the southern wolves (classified as the subspecies *Canis lupus pallipes* and *C. l. arabs*) is dotted. In the region of Southwest and South Asia, there seems to be a zone of contact or transition (cross-hatched) such as is always to be expected in the border regions of populations distinct only at the subspecies level. It is difficult to assign single animals in this region to one or other population group.

Fig. 3.4. Where there are only a few direct measurements of the brain and body size, from which relative brain size as a measure of the evolutionary level of the brain can be calculated, an indirect estimation is possible. There have been some misunderstandings in determining the brain size of the southern wolf subspecies *Canis lupus pallipes* in recent years after this name had been incorrectly applied to mountain wolves from Afghanistan kept and bred at various places in Germany. These wolves with larger carnassials actually belong to Central Asian populations of the subspecies *C. l. chanco*, though some may come from a mixed zone. Manfred Röhrs and Peter Ebinger collected an extensive series of southern wolf skulls in order to refute the finding of a particularly small brain size for these wolves. However, they erroneously claimed to have attained their objective: a difference of only 3%. A simple calculation error (comparing the cube root of braincase capacities instead of the capacities themselves) prevented appreciation of the correct conclusion – one that confirmed the result they doubted (a true difference of about 9%). Their data were used at this point in the original German edition of this book. This caused Röhrs and Ebinger to renew their efforts to collect a new series of measurements of Israeli and Iranian wolves. The skulls were not examined for dental characters to exclude wolves other than the true *pallipes*, as these areas are in the contact zone of the northern and (contd on p. 40)

Indirect calculation of relative brain size in southern wolves

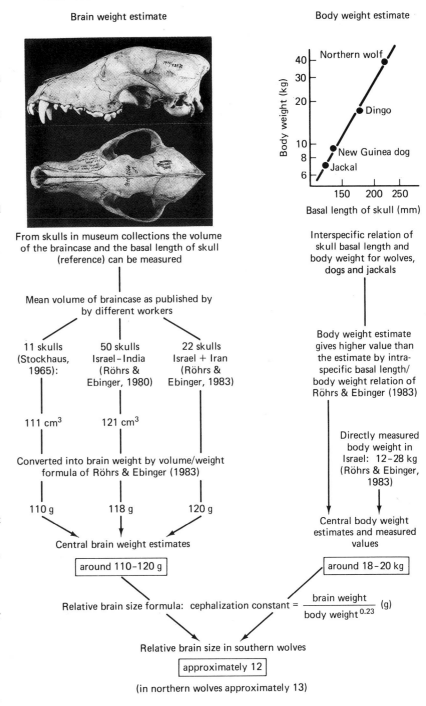

Brain weight estimate

Body weight estimate

From skulls in museum collections the volume of the braincase and the basal length of skull (reference) can be measured

Mean volume of braincase as published by by different workers

| 11 skulls (Stockhaus, 1965): | 50 skulls Israel–India (Röhrs & Ebinger, 1980) | 22 skulls Israel + Iran (Röhrs & Ebinger, 1983) |

$111\ cm^3$ $121\ cm^3$

Converted into brain weight by volume/weight formula of Röhrs & Ebinger (1983)

110 g   118 g   120 g

Central brain weight estimates

around 110–120 g

Interspecific relation of skull basal length and body weight for wolves, dogs and jackals

Body weight estimate gives higher value than the estimate by intra-specific basal length/body weight relation of Röhrs & Ebinger (1983)

Directly measured body weight in Israel: 12–28 kg (Röhrs & Ebinger, 1983)

Central body weight estimates and measured values

around 18–20 kg

Relative brain size formula: cephalization constant $= \dfrac{\text{brain weight}}{\text{body weight}^{0.23}}$ (g)

Relative brain size in southern wolves

approximately 12

(in northern wolves approximately 13)

southern populations as highlighted by the sharply different frequencies of the specific southern wolf pad character in neighbouring populations. Comparing this new mixed series with a collective series of northern wolves that now included the disputed Afghan specimens (which had been shifted between the series instead of excluded as a precaution), they finally produced a statistically insignificant brain size difference no larger than approximately 3%.

comparatively weak carnassial teeth (Fig. 3.5). Such wolves were also distributed throughout Europe in the first stage of the last Ice Age, until about 500 000 years ago. Fossil finds from this period are directly comparable with the modern southern wolves. The wolves of the Central Asian steppes and deserts, the Central Asian mountains and of East Asia have characteristics midway between the southern wolves and the northern wolves of Europe, northern Asia and America, with a larger brain and enlarged, strong carnassial teeth, which makes them appear to have evolved in different directions and to different degrees away from the primitive level.

Comparing characteristics of primitive breeds of dog with these different subspecies of wolf enables the ancestry of the dog to be narrowed down to quite definite geographically separate populations of wolves. The dog differs from most subspecies of wolf and the rest of the species of the genus *Canis* in the form of the coronoid process of the lower jaw. In most *Canis*, this process is usually broad and more or less uniformly rounded at the top, whereas in the dog it is more slender and curved towards the rear with a noticeable overhang (Fig. 3.5); such a form occurs in the Iberian peninsular wolves, as well as in many of those of Central, East and South Asia. Moreover, in view of the structure of their dentition, very primitive feral dog breeds such as the Australian dingo (Fig. 2.5, p. 76) can be derived only from this primitive evolutionary grade of wolf, since their carnassial teeth are not enlarged, while the adjacent teeth seem to be comparatively large. Such primitive wolf populations survived beyond the Ice Age only in the Arabian Peninsula and South Asia. In their external appearance these southern wolves display many more resemblances to the dog than do the northern wolves (Fig. 3.6). Their facial expression, with large, round eyes, is more dog-like than wolf-like. As far as their behaviour is concerned, the vocalization, with a higher proportion of short, sharp barking (above all when pleased) than that of the northern wolf, is emphasized. Social structures are probably less marked amongst them than amongst northern wolves. They tend to hunt alone or to live in pairs or at the most small groups, in which respect they are similar to the coyotes but not to the northern wolves. Nevertheless, there is some indication that Arabian wolves tolerate more crowding in large captive zoo groups. For the time being, details are still unknown because no intensive comparative behavioural observations under controlled conditions have been carried out on southern wolves. In Israel these desert wolves are opportunistic feeders, not being dependent as are the social hunting wolves on a few species of wildlife or large livestock. They are often seen scavenging on carcasses and at rubbish dumps.

The results of immunological studies comparing the dog, the North American wolf and the coyote seem to be of special significance in judging the descent of the dog. They indicate that the dog's relationship

Fig. 3.5. Clear differences are to be found between the lower jaws of northern wolves, southern wolves and dogs. Jaws of a northern wolf (top), a dingo as a representative of primitive dogs (second from top) and a southern wolf (third from top) scaled to the same size illustrate this particularly well when their outlines are superimposed on each other (bottom: premolars and molars except for the carnassial tooth, omitted for clarity). The northern wolf (heavy line in the bottom figure) has a stronger (longer and also somewhat higher) carnassial than the southern wolf and the dog; the coronoid process (the process pointing upwards above the joint) of its lower jaw appears mostly broader and evenly rounded, only seldom having a concavity at the rear as in the dog and frequently in the southern wolf (and in the Central Asian wolf, which is otherwise like the northern wolf).

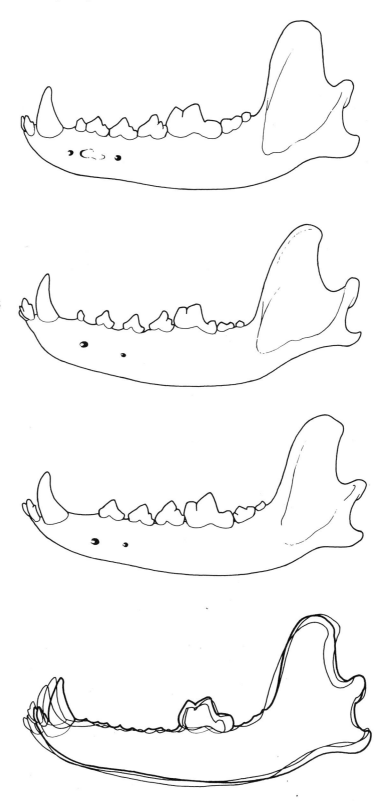

Fig. 3.6. Portraits of northern wolf, southern wolf and primitive dogs. Above left: northern wolf (North America). Below left: dingo (feral Australian primitive dog). Above right: southern wolf (Israel, drawing from a photo in Harrison, D.L. (1968), *The Mammals of Arabia*, vol. II. London: Benn.) Below right: Madagascan primitive dog.

Note the resemblance between the dog and the southern wolf which, in contrast to the northern wolf, is not immediately recognizable as a wolf to the untrained eye.

to the northern wolf is just as distant as its relationship to the coyote. The dog would therefore have to be traced back to a form of wolf that did not participate in the long evolutionary pathway leading to the northern wolf but was subject to a separate development beginning at a very early period not much later than the phylogenetic separation of the wolf and coyote lines (Fig. 3.7). This requirement is fulfilled by the existence of the southern wolves, which, even today, still embody the earliest evolutionary stage of the wolf.

So, from the above, only these primitive wolves can be considered to be the first ancestors of the dog. Nevertheless, perhaps only a marginal local population of such southern desert wolves in the Near East is involved, as their central populations usually have the pads of the third and fourth toes connected from behind, a characteristic not shared by most primitive dogs. The frequency of this pad connection may be as high as 100% in Indian wolves (subsp. *pallipes*) and 80–90% in western Arabian peninsular wolves (Israel desert subsp. *pallipes* and *arabs*); however, fewer

than 20% of so-called Mediterranean *pallipes* wolves of northern Israel, which are more like northern wolves, show this characteristic. This indicates a clear biological separation, in the border zone, of the southern desert wolf populations from the northern ones, pointing to a long period of evolution under relative isolation. This narrow strip permitting crossbreeding with the northern wolf populations gives rise to the special mosaic combination of characteristics that is typical for primitive dogs. Northern wolves must have participated only secondarily in the domestication process by being crossbred with the dog as it spread from its origins in the south of Eurasia towards the north over Europe, North Asia and North America. Also, we cannot rule out the possibility of further genetic enrichment of local dog populations having occurred sporadically here and there as a result of hybridization with the closely related species common jackal and coyote over the course of the millennia. Such crossbreeding produces fully fertile offspring in captivity (Fig. 3.8), but there is no evidence available for natural hybridization with jackals. There are a number of indications of occasional crossbreeding with coyotes from North America but they do not provide conclusive proof.

Very early finds of dog remains that can be distinguished without doubt from those of wolves date from the ninth millennium BC in North America and from the eighth millennium BC in Europe. Because of their geographical location, the American finds clearly indicate that the domestication must have occurred considerably earlier, as the Bering Bridge between East Siberia and Alaska was passable by land, permitting free passage from North Asia to North America for humans and dogs for the last time at the end of the last Ice Age. Canid remains supposed to be

Fig. 3.7. Diagram of the phylogenetic relationships of the species and the progressive brain evolution in the genus *Canis*. The jackals (*C. aureus, C. adustus, C. mesomelas*) embody the most primitive stage of the genus; coyote (*C. latrans*) and southern wolf (*C. lupus, pallipes* group) represent two further evolved intermediate levels, and the northern wolf (*C. lupus, lupus* group) the peak stage.

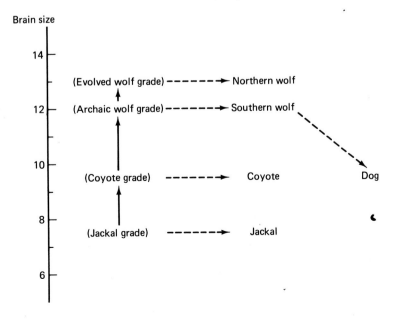

of dogs have indeed been reported from deposits in Alaska that are at least 20 000 years old. Fossil remains from the end of the Ice Age that can be assigned to the dog with some certainty are available from deposits in Japan and also from Iraq (dated at about 12 000 years old). The skeleton of a young dog buried with a human in northern Israel dates to between 12 000 and 10 000 years ago. It is rather more questionable whether further remains from this time from Central and East Europe and Siberia come from dogs or wolves; genuine early primitive dogs may be concealed amongst them. In any case, all these finds show that the dog is the very oldest domestic animal and the only one that came into being at the end of the Palaeolithic at the latest and spread out with the hordes of hunters of that time.

Finally, the enigma of the Falkland Island wolf (*Dusicyon australis*) pointed out by Juliet Clutton-Brock should be mentioned. This animal, assigned to the South American canid genus *Dusicyon* by most of its diagnostic features, had a pelage colour like that of the dingo and also some cranial and teeth features similar to those of the Australian feral dog but sharply contrasting with the smaller mainland *Dusicyon* species. Several members of this species were obviously tamed by the Indians up until the eighteenth and even nineteenth centuries. It cannot be ruled out that the Falkland Island wolf was really a feral hybrid evolved from a cross between introduced, dingo-like primitive domestic dogs and a *Dusicyon* species taken to the Falklands by long-gone human immigrants. Unfortunately, this animal was exterminated in about 1880.

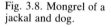

Fig. 3.8. Mongrel of a jackal and dog.

The cat

The closest wild relatives of the cat are in the genus *Felis* of the family Felidae. This genus includes the wild cat (*F. silvestris*), which is distributed from Europe to Central Asia and from South Asia to Africa in three population groups quite different from each other (Fig. 3.9). These three groups used to be, and sometimes still are, classified as three separate species: the European wild cats, the *silvestris* group (Fig. 3.10); the wild cats of Asia, the *ornata* group (Fig. 3.11); and the wild cats of Africa and the Arabian Peninsula, the *lybica* group (Fig. 3.12). Other

Fig. 3.9. Schematic depiction of the former distribution of the wild cat (*Felis silvestris*). Black, European wild cat (*silvestris* group); dotted, African wild cat (*lybica* group); hatched, Asian wild cat (*ornata* group). A contact and mixture zone in the Near East with unclear distribution is bounded by the three groups.

Fig. 3.10. European wild cat (*Felis silvestris*, *silvestris* group; Caucasus population).

members of the genus *Felis* are: the sand cat (*F. margarita*) from the dunes of the deserts of North Africa, the Arabian Peninsula, Pakistan and Central Asia between the Caspian Sea and the western part of the high mountain chains of Inner Asia (Fig. 3.11); the jungle cat (*F. chaus*) distributed throughout the eastern Mediterranean to Southeast Asia; the Chinese desert cat (*F. bieti*) from the southern border area of the Gobi Desert; and the black-footed cat (*F. nigripes*) from the dry areas of South Africa. In all studies of feature complexes, it has been found that amongst all these wild species the domestic cat always closely matches only the

Fig. 3.11. Sitting at the rear, a South Asian wild cat (*Felis silvestris*, *ornata* group); in front for comparison three sand cats (*F. margarita*), a related species from the sandy deserts of North Africa, Arabia, Southwest Asia (West Pakistan) and Soviet Central Asia that was for a time erroneously regarded as an ancestor of the Persian domestic cat.

Fig. 3.12. African wild cats (*Felis silvestris*, *lybica* group). In contrast to the European wild cat, they have no sharp dorsal stripe down the middle of the back but rather an indistinct looking darkening (n.b. animal on the left) and the tabby coat pattern also corresponds in other respects to the striped tabby domestic cat, with striping or spots arranged in stripes and partly run together (n.b. young animal in the centre). The tail is not thick and bushy like that of the European wild cat but thin, like that of the domestic cat.

true wild cat, and of the wild cats not the European wild cat but the African/Arabian wild cat. Some African wild cats look so much like domestic cats that they could be mistaken for them (Fig. 3.12). The diverse coat patterns of these African/Arabian wild cats, namely spots, stripes of run-together spots and extensive pattern reduction, are also to be found in the domestic cat, where they have become the basic patterns of different breeds: the more or less complete striping (striped tabby) is the characteristic of most wild-coloured domestic cats (Fig. 3.13), spots that of the 'Egyptian mau', and the absence of pattern that of the Abyssinian. The general darkening of the colour on the back without formation of a distinct dorsal stripe in African and Asian wild cats is also regularly found in the domestic cat but not in the European wild cat. The original shortness of the hair of the domestic cat is the same as that of the African wild cat and southern Asian wild cat. The same is true of the cranial structure which resembles that of the African/Asian wild cats but not that of the European wild cat in its most important characteristics. The rapid development of the kittens during the first two weeks of their life also seems to be a characteristic of the African wild cat, contrasting with the delayed development of the European wild cat.

It has often been suggested that other wild species participated in the formation of various breeds of domestic cat. In fact, however, there is no sound evidence that interbreeding with any species other than *F. silvestris* could have played a part. Repeated admixture with local wild cats in Europe and Asia is all that can be assumed after the domestic cats descended from the African/Arabian wild cat had spread through the areas where the latter occurred.

Fig. 3.13. Wild-coloured (striped tabby) domestic cat amongst remains of classical Greek civilization with which they first reached European soil as far as is known at present (Athens).

For a while, it was thought that the Siamese cat may have originated by crossbreeding with the leopard cat (*Prionailurus bengalensis*) from South, Southeast and East Asia (Fig. 3.14). The grounds cited for this were a somewhat longer gestation period of the Siamese in comparison with other domestic cats, their loud and somewhat different sounding voice, their narrower skull, a hairless transitional zone between the rhinarium and the bridge of the nose, and a 'phantom pattern' of dots and stripes in 'old-fashioned' Siamese. Also the male is allowed near the

Fig. 3.14. Leopard cat (*Prionailurus bengalensis*), a cat species from South, Southeast and East Asia, for a time erroneously thought to be an ancestor of the Siamese cat.

Fig. 3.15. Pallas' cat (*Felis (Otocolobus) manul*), a species of cat from Southwest and Central Asia, for a time erroneously thought to be an ancestor of the Persian cat.

young and may even take part in their rearing, and the animals have a particularly lively temperament. None of these points is applicable on closer inspection. A longer gestation period correlates with a larger number of young than in the leopard cat. Wild cats sometimes also have the hair trait mentioned, the narrower skull having nothing to do with the essential cranial structure of the leopard cat but rather is the result of the somewhat smaller brain of the Siamese. The 'phantom pattern' appears in the form of the general domestic cat tabby pattern typical of the African wild cat and all special behavioural traits are decidedly ones of the African wild cat. Siamese cats are in reality not like leopard cats, but are even more similar to African wild cats than are European domestic cats.

Reasons were suggested for a participation of either the Pallas' cat (*Felis (Otocolobus) manul*) (Fig. 3.15) or the sand cat (Fig. 3.11) in the Persian cat. Both hypotheses are based on far fewer characteristics than the Siamese cat theory – actually only on that of the long hair, which covers even the pads of the paws – and can also be rejected on closer study. Equally fictional are the role of the jungle cat already mentioned in the origin of the Abyssinian cat and that of the bobcat (*Felis (Lynx) rufus*) in that of the Maine coon cat.

The first records of cat-keeping come from the Old Kingdom of Egypt, from the third millennium BC. The domestic cat is numerous from the New Kingdom in the second millennium BC, when it attained special significance in religious cults as witnessed by many cat mummies and excellent sculptures. Older depictions from the Near East going back to the sixth millennium BC have been interpreted as signs of domestication in the greater Palestinian area even in those times, but these cannot be related unequivocally to the domestic cat. However that may be, there is in any case clear agreement between the cultural history of the domestic cat and the distribution of its ancestral form. Domestication started with populations of the African/Arabian wild cat inhabiting the steppes and deserts from northeast Africa to the west of the Arabian Peninsula.

The ferret

For morphological reasons, two species of the genus *Mustela* of the marten family (Mustelidae), namely the European polecat (*M. putorius*) and the steppe polecat (*M. eversmanni*), can be considered as possible ancestral forms for the ferret (Fig. 3.16). The former is a purely European form, the latter is distributed throughout the steppe regions from eastern Central Europe over the Soviet Central Asian and Siberian area to North America. Skull studies show that only the European polecat can be the ancestral species for the ferret. So far, however, it has been virtually impossible to connect the domestic animal with a particular population of the wild species. There seem to be certain similarities to Spanish polecats in tooth characteristics but knowledge of variations in such characteristics for the areas where the polecat occurs in the Mediterranean countries is still very scanty. In the fourth century BC, Aristotle mentioned that polecats can be tamed. The Romans used ferrets to combat a plague of rabbits on the Balearic Islands, Libyan ferrets being mentioned in this connection. From these few clues, it can be hypothesized that the domestication of the polecat, i.e. the origin of the ferret, took place somewhere in the Mediterranean area in the first millennium BC. Its role as a domestic animal was certainly that of a helper in combatting rodents and rabbits right from the beginning.

Mink and fox

The still undifferentiated definition of a domestic animal given in Chapter 1 would include predatory animals kept on a large scale on farms and subject to selection, particularly for colour variants. Amongst these, the foremost in significance are the mink from the marten family (Mustelidae) and the fox from the dog family (Canidae). American minks (*Mustela vison*) have been bred on ranches since the second half of last century (possibly 1866). Keeping the red fox (*Vulpes vulpes*) on farms began with the breeding of the Canadian silver fox in 1982 and was later extended to include the arctic fox (*Alopex lagopus*). Other carnivores bred in captivity for the production of fur in recent years and decades, though certainly not yet domesticated, are some other species, especially those of the marten and dog families.

The horse

Two species of the horse family (Equidae), the horse and the donkey, have become domestic animals. Wild horses (*Equus ferus*) were widely

Fig. 3.16. Wild-coloured ferret.

distributed throughout Eurasia and North America during the Pleistocene and were important game for hunters in northern latitudes at the end
of the last Ice Age. The geographically separate populations that evolved
in this extensive distribution area in the course of the Ice Age are in many
respects so different from each other that they were regarded for a long
time, and by some still so today, as a group of closely related but separate
species. The only wild horse to have survived is the Przewalski horse
(*Equus ferus przewalskii*) from the Gobi region in Central Asia (Fig.
3.18). There are at most only a few small herds remaining in the wild, if
any at all, but its preservation in zoos seems to be assured thanks to the
constant attention paid to it since it was first brought into captivity and
due to the maintenance of an international studbook.

Wild horses were still very frequent in Central and East Europe in
Roman times. The Roman author Pliny wrote of large herds of wild
horses in the north in the first century AD. They died out in Central
Europe during the Middle Ages, while they survived into the eighteenth
century in the northeast of Poland and in Lithuania. The last ones were
kept in captivity in the animal park of the Polish Prince Zamoyski until
this herd was disbanded at the turn of the year 1812/1813, the horses being
distributed amongst the peasants in the neighbourhood. In the steppes of
the Ukraine, the wild horses (known there as tarpans) were still frequent
in the seventeenth century, but then their numbers were reduced by
constant hunting (Fig. 3.17). It must be assumed that there was already a
mixture of domestic horse in the last free-living tarpans after it had
become increasingly difficult for animals seeking the companionship of
conspecifics to maintain contact with wild horses alone. The extinction of
the species could no longer be prevented in the second half of the
nineteenth century. A few surviving horses came into captivity, where
they died in 1863, 1879, 1887 and 1918/19, respectively. The only remains
of these today are a skeleton in Leningrad and a skull in Moscow.

The last Polish tarpans handed over to the peasants in the environs of
Zamość were absorbed into the local rural breed, the konik, which itself
had remained quite primitive. Many horses with characteristics of wild
species continued on into the middle of the present century in this region.
Tadeusz Vetulani purchased animals with particularly high resemblance
to tarpans from amongst these and subjected them to a selection pro-
gramme carried out in the Institute for Genetics and Animal Breeding of
the Polish Academy of Sciences in Popielno in the Mazurian Lakeland.
By means of intense selection on the basis of the konik he managed to
reproduce horses that embody the external appearance of the tarpans in a
uniform way, with the exception of the overlong mane and tail (Figs. 3.17
and 3.19).

The tarpan and Przewalski horses are quite distinctly different in
several characteristics. While the latter has a heavy head with a long
muzzle and large teeth, the East European wild horse was characterized

Fig. 3.17. Portraits of wild and domestic horses. Above left: Przewalski horse. Below left: example of a medium-sized European breed of domestic horse, the Haflinger. Above right: foal of a tarpan, the extinct Eastern European wild horse (reconstruction drawn from nineteenth century depictions). Below right: Eastern European primitive horse: Polish konik, a breed of tarpan-like animals.

Note the heavier effect of the head of the Przewalski horse due to its longer muzzle (relation of the distances eye to ear and eye to upper lip), and the upright mane of the wild horses.

by a shorter muzzle and weaker dentition, and by a slender body with muscular legs (Figs. 3.17–3.19). In addition to the dun, pale brownish colour of the Gobi wild horse, the tarpan exhibited mouse-grey tones, such as now are preserved in the konik back-crosses. Primitive breeds of domestic horse are much more reminiscent of the tarpan than of the Przewalski horse. Finally the different chromosome number of the Przewalski horse supplies the decisive clue that it can in fact have played hardly any part in the origin of the domestic horse because, in contrast to all domestic horses, which have a diploid set of 64 chromosomes, the Mongolian wild horse has a set of 66.

It is extremely difficult to ascertain from the subfossil bone finds when the primary domestication of the horse took place, since there is no known characteristic which would allow primitive domestic horses to be distinguished with certainty from European wild horses in the skull or in other parts of the skeleton. Reliable identifications are therefore first possible from the time when the character of the domestic horse is confirmed by written or pictorial records. For earlier times, the only possibilities are guesses based on altered frequency ratios and on the age

distribution of horse bones. First indications of the domestic horse are found in the Orient in the third millennium BC; the further spread to lands where there were no wild horses was apparently only at the beginning of the second millennium BC. Domestication seems to have occurred in the steppes north of the Black Sea and the Caspian towards the end of the fourth millennium BC, so that the domestic horse does indeed seem to have originated from the tarpan of eastern Europe.

Fig. 3.18. Przewalski horses.

Fig. 3.19. Tarpan-coloured East European horse: selection for tarpan characters from Polish koniks.

The ass

Wild asses (*Equus africanus*) are today distributed only in small residual stocks in the northeast African area from Nubia to Somalia, although they were also found in the Atlas area of North Africa in historical times. Whether the Arabian Peninsula was also inhabited by these strongly desert-adapted animals is still unknown, although it is sometimes mooted. There are distinct differences between the various regional populations in the colouring, striping, body size and skull characteristics. The Nubian wild ass has a short, broad – or long, thin – black band across the shoulders to the upper foreleg that forms a shoulder cross with the

Fig. 3.20. Nubian wild ass.

Fig. 3.21. Compared to the Nubian wild ass, the wild-coloured domestic ass has a very strong shoulder stripe.

black stripe along the middle of the back (Fig. 3.20). Its legs are hardly ever striped. The larger Somalian wild ass is characterized by heavy striping of the limbs but has either no shoulder cross or only a very weak one. Its colouring tends more to the yellowish-reddish than the Nubian wild ass, which is more grey. The Atlas wild ass must have had stripes on its legs as well as a shoulder cross, to judge from ancient artistic representations.

As a rule, the domestic ass also has both types of marking, but its black cross stripe is often much broader and longer than that of the known wild forms (Fig. 3.21). There is, however, considerable variation, ranging from the absence of this band, via a very thin marking, to a long, broad band that even has a forked lower end. Apart from deviations into dark brown, black or white not found in the wild animals, their colouring includes both the grey and the more reddish-sandy shades of the various wild forms. So it can probably be assumed that, as with other breeds of domestic animals, secondary crossbreeding has taken place with populations other than the one with which domestication began. The archaeological finds indicate a primary domestication in the Nile Valley in the fourth millennium BC and a subsequent spread through Palestine at the beginning of the third millennium BC.

The pig

Wild pigs of the genus *Sus* (family Suidae), amongst which the ancestors of the domestic pig are to be sought, inhabit Europe, practically all of Asia, and North Africa. The largest part of this huge area is inhabited by the wild pig (*S. scrofa*: Fig. 3.22). There are extensive differences

Fig. 3.22. Schematic depiction of the distribution of the wild pig (*Sus scrofa*). Black, *scrofa* group; hatched, South Asian *cristatus* group; closely dotted, Indonesian *vittatus* group. Areas where the black resolves into dots: in West Asia, conjectured but not more exactly known contact or transition zone from the *scrofa* to the *cristatus* group; in East Asia, transition from more *scrofa*-like to more *vittatus*-like populations with similarly still largely unknown transition to the *cristatus* group.

Fig. 3.23. Pig portraits.
Right: adult warty pig
boar (*Sus celebensis*) from
Sulawesi (Celebes) (from a
photo in Mohr, E. (1960),
Wilde Schweine.
Wittenberg: Ziemsen).
Left: banded pig (*S. scrofa
vittatus*) (from a photo in
Mohr, E., *ibid.*) Below:
Polynesian pig (mixed
population originating in
animals from the circle of
the Papua pigs and
European–East Asian
pigs).

between the various geographically separate populations of this species. The most primitive form with respect to general evolutionary level, expressed mainly in progressive cephalization, is the banded pig (*S. scrofa, vittatus* group: Fig. 3.23) of Southeast Asia. It is characterized by the smallest relative brain size within the species (Fig. 3.24) and has comparatively non-specialized teeth and an equally non-specialized cranial structure. Its distribution area extends from Sumatra and Java in the Indonesian region to some of the Lesser Sunda Islands. Through China, Japan and East Siberia, there is a gradual transition from the banded pig into the eastern populations of the true wild pig (*S. scrofa, scrofa* group), which has a much larger brain and longer snout with stronger teeth. Another transition leads away from the banded pig

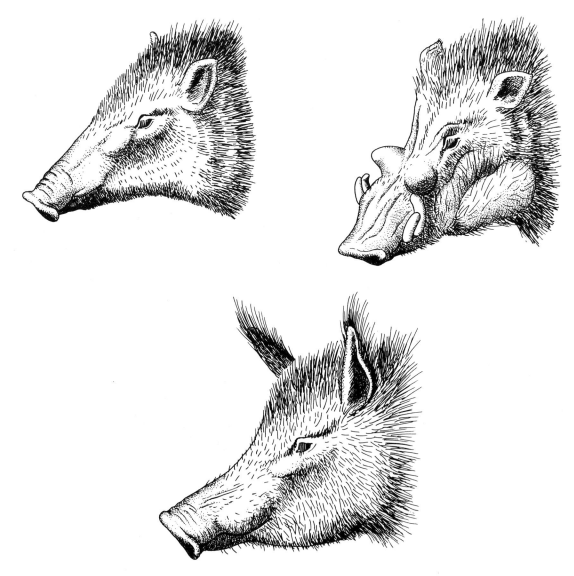

towards the Indian wild boar (*S. scrofa, cristatus* group), going west in South Asia. This form has a somewhat increased brain size, though not so much as in the *scrofa* group, and a very characteristic, particularly large and complex last molar.

The warty pigs (*verrucosus* group: *S. verrucosus* and *S. celebensis*) are close to the banded pig in their primitive evolutionary status. The special external features that give them their name are three pairs of prominent facial warts on the snout in front of the eyes, on the cheeks below the eyes and on the angle of the jaw, which can be small or even absent in sows and huge in old boars (Fig. 3.23). The skull has much more robust bones with heavier superstructures than that of the banded pig; the brain size corresponds approximately to that of the banded pig (Figs. 3.24 and 3.25). The warty pigs of today are distributed throughout Java, Sulawesi, the Lesser Sunda Islands, the Moluccas and the Philippines. At the beginning of the Pleistocene, they were also to be found in Europe and so were a genuine old evolutionary grade of wild pig before they were replaced by pigs of the *scrofa* group in the Cromer Interglacial. Other related species are the bearded pig (*S. barbatus*), with slender, long legs and a beard, from Malaysia, Sumatra, Borneo, the Philippines and some smaller islands, and the diminutive pygmy hog (*S. salvanius*) that used to inhabit the east of the Himalaya region and is now probably extant only in small numbers in the Manas national park in Assam.

The most primitive of the modern domestic pigs seem to be the Papua pigs that live mostly semi-ferally or ferally in New Guinea and the surrounding islands and are also spread through the region of the Pacific islands in the form of Polynesian pigs (Figs. 3.23, 3.25 and 9.14 (p. 140)). Any colonization of New Guinea by pigs other than those introduced there by human agency can in all probability be ruled out, since no other wild-living larger mammal of a late Tertiary or later evolutionary level managed to cross the sea straits on its own to this island originally inhabited only by marsupials of the Australian region. These pigs in New

Fig. 3.24. Diagram showing the relative brain sizes of the warty pig (*Sus verrucosus* group, filled circles), banded pig (*S. scrofa, vittatus* group, plus signs) and Middle Eastern wild pigs (*S. scrofa, scrofa* group, open circles), according to skull measurements (cranial capacity instead of brain weight; basicranial axis as the reference measurement).

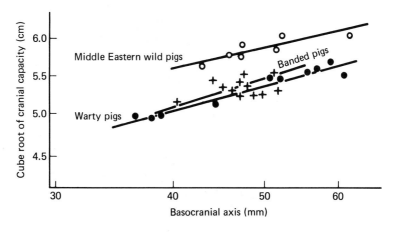

Guinea possess a mosaic of characteristics of the warty pig and banded pig, which, taken as a whole, makes them seem closer to the warty pig. The pigs on the island of Timor, most likely introduced, even seem to be entirely of the warty pig type. The Sulawesi warty pig still exists as a domesticated form on the island of Roti near Timor even today, occurring as a black or black and white type. Single warty pig characteristics are indicated even in improved breeds of Chinese domestic pig. So the root of pig domestication was not only in the wild pig, as always used to be supposed, but certainly also in the Sulawesi warty pig. Some skull and teeth characteristics of primitive breeds of domestic pig point to the banded pig as the second, i.e. wild pig, root of the domestic pig. Accordingly, Colin Groves assumes, probably correctly, that pig domestication began with warty pigs transported from Sulawesi to the neighbouring islands and that these were further crossbred with banded pigs, which had likewise been imported, resulting at first in the Papua pig.

Fig. 3.25. Skull characteristics of the warty pig (above right) alongside Papua pigs from New Guinea. Compared with the skulls of wild pig boars (*Sus scrofa*), the skull of warty pig boars (*S. celebensis* and *S. verrucosus*) has heavier bone, with much stronger relief-like modelling of the surfaces in front of the eye sockets (thin arrows), at the lower edge of the zygomatic arch (black star) and at the back lower edge of the lower jaw (white star). There is a large bulge (thick, white arrow) on the side of the lower jaw caused by the root of the canine tooth. These characteristics are found to match partly completely and partly in attenuated form, skulls of old boars of the Papua pig (upper skulls at the right middle and bottom, lower jaws above and below left).

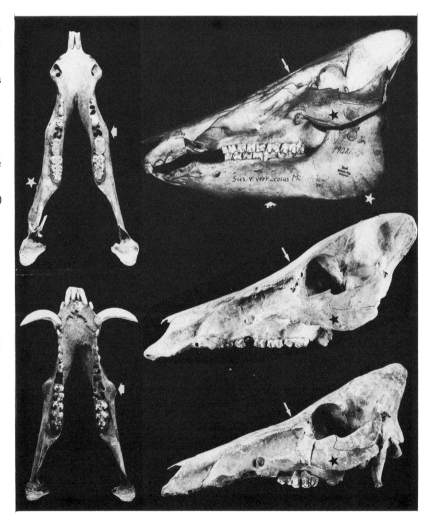

The oldest indications for the use of the pig on New Guinea go back about 10 000 years, at which time the people of that region had apparently brought pigs with them over the sea. It was only some 1000 years later that the oldest domestic pigs were to be found in the region of Asia Minor and the Near East, the starting point of their further spread to Europe. In the course of the Neolithic, as mentioned in the previous chapter, secondary domestications, which rapidly multiplied stocks of domestic pigs and shifted their appearance more towards that of the European wild pig, occurred in Europe as a result of considerable catches of wild pigs and subsequent crossbreeding. This, however, did not go so far as to alter the characteristic chromosome number of the domestic pig from 38 to the 36 of the Central European wild pigs. The Southeast European and Asian wild pigs have the same chromosome number as the domestic pig.

The Old World camels

In the desert belt of the Old World, two domestic species of the camel family (Camelidae) play a decisive role in human life: the dromedary (or single-humped camel) in the west, from the Sahara via the Arabian Peninsula to the Near East and India, and the Bactrian (or two-humped camel) in the east, from the deserts of the Near East to Mongolia. Wild camels (*Camelus ferus*) now exist only in the Gobi Desert. These two-humped camels are probably members of the ancestral species of the domestic camel and, according to fossil finds from the Pleistocene and from Postglacial deposits, used to be distributed at least as far as the Soviet Central Asian region between the Caspian Sea and the western chains of the Central Asian mountains. This former incidence of wild camels makes it difficult to interpret whether the earliest evidence of camels from the Near East comes from wild or domestic animals. The first clear evidence of a domestic status dates from the third millennium BC in the Turkmenistan and eastern Iran area.

In contrast to the two-humped camels, the wild precursor of the dromedary is still quite unknown to us. The genus *Camelus* was certainly originally distributed throughout the whole desert and desert–steppe area from the Sahara to the Gobi; fossil records from North Africa, however, indicate that the form still extant there in the Postglacial period, *Camelus thomasi*, resembled the two-humped camel rather than the dromedary, so that it can be excluded from the search for the wild dromedary. The only remaining area for the distribution of the wild dromedary is therefore the Arabian Peninsula. Roman authors report that there were wild camels in Arabia. This region also appears in the written records of the early civilized peoples of the Near East and in all available archaeozoological finds as the centre from which the domesticated dromedary spread, so that a domestication there towards the end of the fourth millennium BC can be proposed.

The New World camels

The small South American camels, llama and alpaca, are the only large domestic animals to have originated in South American Indian civilizations. The alpaca inhabits the Andes of Peru and Bolivia; the llama is bred both there and in the northwest of Argentina and the north of Chile. In the nineteenth century, it was still to be found also in Ecuador and Paraguay. The possible wild ancestors in the same areas are the guanaco (*Lama guanikoe*) (Fig. 3.26) and the vicuna (*Lama vicugna*) (Fig. 3.28). It is very difficult to trace the course of domestication in detail in the available archaeozoological finds, due to the problems of distinguishing between the remains of wild and domestic animals. There is evidence of guanaco and vicuna hunting in the central Peruvian puna (the tableland at over 4000 m) which dates to 7000–4000 BC. In the bone collections, alpaca-type incisors indicate the early existence, in the fourth millennium BC, of at least one of the domestic species. In the succeeding periods, there is a progressive accumulation of evidence of llama and alpaca herding.

Whereas the llama (Fig. 3.27) is a domesticated form of the guanaco in all characteristics (morphology, biochemistry, behaviour), the origin of the alpaca (Fig. 3.29) has been much disputed for a long time. Four theories, which were in part mutually exclusive, have been under consideration. One regarded the alpaca as a domestic form of the vicuna, another as a second domesticated form of the guanaco. The third theory envisaged it as a product of a mixture between vicuna and llama (which are known to be able to breed together), while the fourth sought a third, now extinct, wild species intermediate between guanaco and vicuna as the alpaca's ancestor. Recently, after the study of numerous complexes

Fig. 3.26. Guanaco, the wild ancestor of the llama.

of characteristics of these species, it has become easier to decide between these hypotheses.

The great variability in cranial structure makes it difficult to relate the alpaca to any one of the wild species. Only one characteristic gives it greater resemblance to the guanaco or to the llama. The structures of the lower incisors of the wild species, the guanaco and the vicuna are distinctly different. The vicuna's incisors wear down more quickly because of the special conditions in its habitat; they have open roots and so can grow continuously. They have an enamel layer only on the outer

Fig. 3.27. Llama, the domestic form of the guanaco.

Fig. 3.28. Vicuna, the second species of South American wild camelid. Along with the llama, it belongs to the ancestry of the alpaca.

side, whereas the teeth of the guanaco and llama have a uniform enamel layer on both the outer and the inner side. Alpaca incisors are intermediate between these two forms: the roots range from closed to permanently open with longer lasting growth. The enamel covering is thicker on the outer than on the inner side. A similar intermediate position is found in the form of the metatarsal callosity. It is clearly visible in the guanaco and llama, while it is covered with hair from both sides in the vicuna. In the alpaca this covering occurs only from the outer side. The skin thickness of the alpaca is similar to that of the guanaco and llama. Electrophoretic separation of serum proteins reveals a great resemblance between the alpaca and the vicuna but not between the alpaca and the guanaco or the alpaca and llama. The muscle proteins are composed of 20 different amino acids in the vicuna but of 24 in the guanaco and llama. Alpaca muscle also has 24 but the four corresponding to those missing in the vicuna are present in only low concentrations. This intermediate position is also displayed in the behaviour, the vocalization of the alpaca being more like that of the vicuna. When excited, guanacos and llamas raise their tails in a curve like a sickle but the vicuna raises its tail straight up. Both positions, as well as intermediate ones, are found in the alpaca. Further resemblances to the vicuna can be observed in bathing behaviour and kicking, while defecation behaviour is again of an intermediate type.

On the basis of these result, the early theories that the alpaca descends either from the guanaco or the llama alone, or from the vicuna alone, have to be rejected, since in all complexes of characteristics it posseses on the one hand traits characteristic of the vicuna but absent in the guanaco and llama, and on the other traits that are not found in the vicuna but are typical of the guanaco or llama. The theory of a special, extinct ancestral

Fig. 3.29. Alpaca.

species can also be ruled out, since it has been found that the skeletal remains taken as evidence for this theory may really come from the vicuna and there are no other indications of any kind that such a third species ever existed. The only remaining hypothesis is that the alpaca is a mixture of both lines, by crossbreeding of captured vicunas with the only initially available domestic animal, the llama. Such a process parallels both the crossbreeding of local wild animals with populations of related domestic animals and the interbreeding of closely related domestic animals such as the horse and the ass, or the dromedary and the Bactrian camel.

The reindeer

The reindeer (Fig. 3.30) was the only domestic animal from the deer family (Cervidae), until very recently (see Chapter 12). Since the Palaeolithic, when the humans of the last Ice Age colonized the cold steppes and tundras up to the ice in the north, the reindeer (*Rangifer tarandus*) has been an important, and later even the only, food source in the north of Eurasia. It was there that domestication, the time course of which is unknown, took place. It is possible that the beginning of domestication goes back to about 15 000 years ago, i.e. the end of the Palaeolithic. This transformation in use from a game animal into a domestic animal was completed by the first millennium BC at the latest. In North America, the reindeer (caribou) remained undomesticated.

Reindeer-keeping is still at an early stage of domestication in many areas. Crossbreeding with wild reindeer from local populations is frequent. The Chukchi in northeast Siberia often use their half-tamed

Fig. 3.30. Domestic reindeer stag.

domestic reindeer to catch wild ones, which are used chiefly as a meat supply. Small groups of wild reindeer join large domestic herds fairly readily but the reverse is also possible in that small groups of domestic reindeer join up with larger wild groups and disappear. Apparently the herd stability, the cohesion of the animals, increases as does the size of the herd, and the tendency of the domestic animal to return to the wild decreases correspondingly. Both returning to the wild and capturing wild animals with the aid of domestic reindeer seem to be a function mainly of the herd size, on the basis of the fundamental gregariousness of the species. Consequently, a certain constant genetic flow between wild and domestic animals is still possible in these regions.

Cattle

Genuine cattle form a complex of four genera within the large family of bovine ruminants (Bovidae). The most primitive form phylogenetically seems to be the small anoa (*Bubalus depressicornis*) from Sulawesi. It is related to the large Asian water buffalo (*B. arnee*), which is closely allied to a smaller form, more like the anoa, on the island of Mindoro in the Philippines. The buffalo group is represented in Africa by the African buffalo (*Syncerus caffer*). The second group of genuine cattle is formed by the bison, which inhabited Eurasia and North America in a large number of forms during the Pleistocene. The three forms still extant today are only a small residue. These are the European bison or wisent, the Canadian wood bison and the North American plains bison. As a rule, the latter two (*Bison bison* with the subspecies *bison* and *athabascae*) are classified as one species with the wisent (*Bison bonasus*) as another separate species. However, since the wood bison is a form intermediate between the wisent and the plains bison, and all three interbreed easily when the geographical barriers are removed, it would be equally justifiable to treat them either as three separate species or as subspecies of a common Euro-American species. The largest diversity of species is to be found amongst the wild cattle in the narrow sense. Of these, the aurochs or urus (*Bos primigenius*) died out in the greater part of its distribution area (Europe, North Africa, the Near East and South Asia) quite early. In Europe it survived to the Middle Ages. The last herd left was preserved in Poland and consisted of 38 animals in 1564. By 1599 there were only 24 left, in 1602 only 4, and the last cow died in 1627 (Fig. 3.31).

The wild yak (*Bos mutus*), which is particularly close to the bison according to recent findings of Colin Groves, inhabits the desert steppes of the North Tibetan tableland. The gaur (*Bos gaurus*) is found in South and Southeast Asia. The Southeast Asian region is the habitat of a further wild bovine, the banteng (*Bos javanicus*), which is also to be found on the islands of Java and Borneo (Fig. 3.32). That the kouprey (*Bos sauveli*) from Indochina is an independent wild species has been disputed, but as yet too little is known about it. The kouprey has some mixture of banteng and zebu features in its appearance and cranial characteristics, and its skull is said by Colin Groves to be close to that of the aurochs.

Perhaps, apart from this last, uncertain one, forms of domestic cattle can be related to all species of the genus *Bos*, while the only buffalo to have been domesticated is the water buffalo and no domestic animal originated from the genus *Bison*. Taking into account all known characteristics of the aurochs, it is the only wild species that can be regarded as an ancestor for domestic cattle in the real sense. However, insufficient knowledge on the characteristic peculiarities of the various geographically separate populations of the aurochs make it largely impossible to

Fig. 3.31. A 'zoo-born' domestic cattle breed embodying the appearance of the extinct aurochs ('urus' in German zoos).

Fig. 3.32. Banteng (*Bos javanicus*), the wild ancestor of the Bali cattle (bull black, cow brown).

relate primitive breeds of domestic cattle to one or other of these populations. The same is true for the origin of the zebu, a complex of cattle breeds that differs distinctly from all other breeds in many characteristics. Zebus are characterized not only by their humps but also by their slenderness and long legs as well as usually large dewlaps. They can bear heat much better than other breeds. They have been called the 'greyhounds' amongst the cattle (Figs. 2.8 (p. 19) and 3.33). Two hypotheses on their origin have been suggested. The first proposes an independent domestication of a Southwest–South Asia form of aurochs. The second explains the origin of humped cattle as due to selection from humpless, long-horned, primitive domestic cattle in the hot steppes at the eastern edge of the Great Iranian Salt Desert. This dates the transformation to the fourth millennium BC, since the typical characteristics are found from the end of the fourth and the beginning of the third millennium BC at the latest. At present, the oldest finds relating to domestic cattle come from strata dating to the seventh millennium BC in the north of Greece and in Anatolia.

A real solution to the problem of the origin of humped cattle will not be possible until more is known about the geographical diversity in the characteristics of the aurochs. Without this knowledge it will scarcely be possible to decide on the region where the primary domestication of the aurochs took place with any certainty. Possible similarities to the kouprey are of the greatest interest in this connection.

The wild yak is the ancestor of the domestic yak, use of which has remained confined to the Inner-Asian region (Fig. 3.34). There are hardly any certain clues as to when its domestication took place in prehistoric times.

Domestic Bali cattle arose from the banteng in the Indonesian islands,

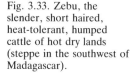

Fig. 3.33. Zebu, the slender, short haired, heat-tolerant, humped cattle of hot dry lands (steppe in the southwest of Madagascar).

though the date is just as unknown as for the domestic yak. Javanese records mention exports of cattle from Bali to Java in the fourteenth century, so that the existence of the domestic animal is certain only from the Middle Ages onward. Secondary domestication continued in this case until modern times, wild bantengs still being caught to be used in breeding Bali cattle in the eighteenth century. The same is so for the mithan or gayal, the domestic form of the gaur, which is found in Assam, Burma and Bhutan. Continual crossing with wild gaurs to improve the mithan breed was welcomed there. Nothing is known about when the primary domestication took place in this case either.

Multiple crossbreeding between these regional south and southeast Asian forms of domestic bovine and domestic cattle of the aurochs type ensured a continual gene flow between the species and the emergence of characteristics from one form in the others. Hybridizations of zebus with Bali cattle and mithan enriched the gene pool and thus the conformation of the two latter forms. A stabilized crossbreed population of Bali cattle and zebus is found on the Indonesian island of Madura (Madura cattle). Such hybridizations give rise to the idea that the zebu may have provided the original impetus for domestic use of the banteng and gaur.

Just as the originally imported cattle stocks were increased by catching aurochs and finally by crossbreeding with them in the European Neolithic (cf. Chapter 2), so it is conceivable that a rapid increase of a small, at first imported stock of zebus was attempted in a similar way, except that the available indigenous wild species were not aurochs but the related species, banteng and gaur. The admixture could have been attempted by capturing wild cattle as well as by mating wild bulls with domestic cows. So it may well be that the origin of the mithan and Bali cattle was not a

Fig. 3.34. Domestic yak, the long-haired, cold-tolerant domestic cattle of the Inner Asian tableland, a product of the domestication of the wild yak.

case of primary, but rather of secondary domestication comparable to the regional origin of primitive breeds of other domestic animals by cross-breeding with the various closest related wild forms locally available. Parallel cases would be: the modern domestic pig resulting from the two roots warty pigs and banded pigs, in addition to numerous later cross-breedings with other wild pigs; the creation of the alpaca by crossbreeding the domestic llama with the wild vicuna; or perhaps also the Falkland Island wolf. In such a case it is fruitless to search for the primary domestication of the banteng and gaur. A similar occurrence is also conceivable for the yak. Crossbreeding domestic cattle and domestic yaks still takes place everywhere in the Central Asian mountain ranges today. These two forms were used in the Altai Mountains to breed a new kind of cattle combining the most important economic features of the two for that region, namely the low sensitivity of the domestic yak to climatic conditions and its usefulness as a beast of burden in mountains combined with the high milk production of the domestic cow. Still another new breed of cattle, the 'cattalo', 'beefalo' or 'American Breed' in Canada and the USA, is based on a similar hybridization of two species. By back-crossing the fertile females from the first generation of a cattle–bison hybridization, a breed was produced that has between half and three-quarters of the characteristics of the domestic animal and that is particularly resistant to cold and so loses less weight than other breeds of cattle on the open, snow-covered ranges in the North American winter.

Cattle can therefore be regarded as a pattern for the domestic use of the local wild animals in each section of the whole European–Asian area in which the stock of domestic animals reflects the relevant regional form of the genus *Bos*. The question which must be posed is whether there were multiple primary domestications, each independent of the others, or secondary domestications catalyzed only by the first domestic cattle – the aurochs' progeny. If this is accepted, then there may be a similar situation for the domestication of the water buffalo from the genus *Bubalus*. However, the crossbreeding of buffalo into stocks of domestic cattle is very difficult due to the low rate of breeding success. Nevertheless, the possibility remains that it was the idea of catching wild cattle in an attempt to increase stocks of domestic cattle that was instigatory in the domestication of buffalo. At present, evidence on the centre and the date of this event is still scarce. The first archaeozoological records of the domestic water buffalo come from the third millennium BC in India and Mesopotamia. The original distribution of the water buffalo covers both these areas so that it is not possible to narrow down the possibilities to a particular centre of domestication in this way either. No further analyses of characteristics are available at present.

The sheep

In addition to genuine cattle, the bovids (Bovidae) have provided two other extraordinarily important domestic animals: the sheep and the goat from the caprine subfamily (Caprinae). The ancestors of the domestic sheep are to be sought among the wild sheep of the genus *Ovis* distributed in a large number of very different geographically separate forms from Europe to North America (Fig. 3.35). The relationships between these populations in view of their taxonomic division into separate species or subspecies is disputed, as is reflected in the different scientific nomenclature for the individual forms. The moufflon (*O. musimon* or *musimon* group of *O. ammon*) originally inhabited Sardinia and Corsica as feral sheep and the eastern Mediterranean area to northwest and south Iran as truly wild sheep. They are comparatively small, short-snouted sheep with broad tails and sometimes a saddle-shaped marking on the back. Moufflons living ferally in Central Europe and other places in the world were imported Sardinian or Corsican animals (Fig. 3.36), which in turn were evidently brought into those islands in prehistoric times, probably from Asia Minor. All regional populations of this group have 54 chromosomes. Further to the east of the Middle East, the urial group (*O. vignei* or *vignei* group of O. ammon) adjoins the moufflon area and extends from northeast Iran via Soviet Central Asia to Tadzhikistan and Afghanistan. They are larger with a thin tail and short snout, and have a diploid chromosome set of 58. The distribution area of the argali (*O. ammon* or *ammon* group of *O. ammon*) extends from the Central Asian mountains of Pamir and Tien Shan in the west to the Chinese mountains in the east and the Altai in the north. Members of this group are very large,

Fig. 3.35. Distribution of wild sheep. Dotted areas in Asia Minor and the Near East, moufflon (*musimon* group with 54 chromosomes); hatched, urials (*vignei* group with 58 chromosomes). Between these two is a transitional zone. Black, argalis (*ammon* group with 56 chromosomes); dotted areas in East Siberia and North America, Amphiberingean wild sheep (with 54 chromosomes at least in the case of the two North American forms).

long-snouted and possess a diploid chromosome number of 56. Finally the sheep on either side of the Bering Sea are medium sized to large and relatively short legged. They differ in several anatomical features from the groups mentioned so far. The group includes the snow sheep (*O. nivicola*) of East Siberia as well as the Dall sheep (*O. dalli*) and the bighorn (*O. canadensis*) of the mountains in the west of North America. These American wild sheep have 54 chromosomes.

In view of their general combination of characteristics, the members of this Northeast Asian and North American group can be excluded from the ancestors of the domestic sheep, all of which have a diploid set of 54 chromosomes. The only origin which can be considered for them is therefore the moufflon group originally from southwest Asia. Biochemical studies agree with this finding. Consequently, the centre of domestication is to be sought west of a line from the Caspian Sea to the Gulf of Oman, which agrees with the archaeozoological data. The earliest finds relating to domestic sheep date from the ninth millennium BC in Southwest Asia, though distinguishing remains of primitive domestic sheep from those of wild ones in areas that the latter inhabit seems to be difficult. This problem of a very slowly proceeding domestication process lasting over many centuries is illustrated by the Corsican and Sardinian moufflon population. These moufflon of the Tyrrhenian are obviously not truly unchanged wild animals, but consist of animals that were introduced in the sixth or even the seventh millennium and returned to the wild state at a very early stage in the domestication process.

The primitive breed of fully domesticated sheep most resembling the moufflon that has survived into the present is the Soay sheep. A feral population of this breed on the Soay island of the St Kilda group, west of the Hebrides off the northwest coast of Scotland, has remained

Fig. 3.36. European moufflon.

Fig. 3.37. Soay sheep, the wild-coloured, primitive breed of domestic sheep most resembling the moufflon.

untouched by all later developments in the breeding of the domestic sheep (Fig. 3.37). Soay sheep origins may go back to European Bronze Age sheep. The first domestic sheep in Europe dated with certainty appeared in the Balkans in the eighth to seventh millennium BC. The dating of sheep finds from Dobruja, first thought to be the end of the Ice Age or early Postglacial period, is subject to doubt.

The goat

As with the wild sheep, there are several forms of wild goat (Fig. 3.38) and it is disputed whether they are taxonomically separate species or extremely different subspecies of a common species. The bezoar goat (*Capra aegagrus* or *aegagrus* group of *C. ibex*) of the eastern Mediterranean area and southwest of Asia to Afghanistan is characterized by sabre-shaped horns with sharp front edges, a dorsal stripe, dark shoulder cross and dark leg markings (Fig. 3.39). Distinctly different from these are the markhor or screw-horn goats (*C. falconeri* or *falconeri* group of *C. ibex*) from the mountains of Afghanistan to Kashmir, Tadzhikistan and Uzbekistan, with corkscrew-shaped horns that give them their name, a mane on the bucks and a lack of sharp contrast in their colouring (Fig. 3.40). There exists a very interesting wild goat population in the Chiltan range, Pakistan, judged to represent an extreme type of the *falconeri* group by some authors and an extreme type of the *aegagrus* group by others. This goat combines bezoar-like colour and lack of a ruff with an open screw-horn of markhor type but bezoar-like anterior keel. A third large group is formed by the ibexes (*C. ibex* or *ibex* group of *C. ibex*),

Fig. 3.38. Distribution of wild goats. Black, ibexes (*ibex* group); dotted, bezoar goats (*aegagrus* group); hatched, markhor goats (*falconeri* group).

whose horns have a broad front surface rather than an edge like that of the bezoar goat, and whose colouring has single elements of the bezoar coat marking in varying degrees (Fig. 3.41). Four subgroups of different geographically separate populations of ibexes can be distinguished: the Spanish ibex of the Iberian mountains, the ibexes of the Alps and Central Asian and Siberian mountains, the Caucasian ibexes or turs, and the

Fig. 3.39. Bezoar goat, ancestral form of the domestic goat. In contrast to the other wild goats, it has a characteristic contrasting colouring such as is also found in wild-coloured domestic goats.

Fig. 3.40. Markhor.

Abyssinian and Nubian ibexes found from Abyssinia to the Arabian Peninsula.

The relation of the domestic goat to the bezoar among these different forms of wild goats seems clear. It corresponds to the bezoar in its wild-type colouring (Fig. 3.42); its horns, except when they are of a special form, mostly have the typical sabre shape of the bezoar. The corkscrew horns frequently found in domestic goats (Fig. 3.43) are twisted in the opposite direction to those of the markhor and so are not related to it. However, markhor-type horns similar to those found in the wild Chiltan goat are present in Cherkessian goats from the regions between the Caucasus and Mongolia. Such screw-shaped horns are also found in

Fig. 3.41. Ibex.

Fig. 3.42. Male domestic goat of wild-coloured type (African dwarf goat), with the colour distribution of the bezoar goat.

depictions from the areas of Ancient Egypt and Ancient Greek civilizations (Fig. 3.44). This may indicate that the Chiltan goat played a part in
goat domestication. The oldest evidence for this process comes from the
distribution area of the bezoar, namely Iran, and goes back into the ninth
millennium BC. It seems therefore that, after the dog, sheep and goats
are the next oldest domestic animals.

Fig. 3.43. Most frequent
form of twisting in the
horns of domestic
billygoats: twisted in the
opposite direction to the
horns of the markhor
(goats from the feral
populations of the
Hawaiian islands).

Fig. 3.44. Reconstruction
of a domestic goat from
Ancient Greece, with
horns twisted in the same
direction as those of the
markhor. From small
bronzes in the National
Museum of Archaeology,
Athens.

The rabbit

The domestic rabbit is clearly a descendant of the European wild rabbit (*Oryctolagus cuniculus*) from the order Lagomorpha. It was originally confined to the Iberian Peninsula in the Postglacial period. The Phoenicians may have been the first people from the early civilizations of the Mediterranean region to have had contact with the wild rabbit about 1100 BC. It first became known in Graeco–Roman areas in the second half of the first millennium BC. Romans then released rabbits on the islands of the western Mediterranean, so initiating their later world-wide spread. Enclosures originally built to keep hares were then also used to keep rabbits in the Roman Empire. Keeping them in captivity must then have led to breeding them in captivity, which by the Middle Ages had spread over western and Central Europe. At that time, large rabbit runs were used for hunting by royal courts, including women. It must have been then that the domestic rabbit arose, kept in more confined quarters alongside the wild rabbits in such runs, a process in which French monasteries probably played a leading role. The first clear records of flourishing domestic rabbit-keeping and breeding, including the existence of various colour forms and sizes, come from the sixteenth century, so that the domestication phase must have been completed by the end of the Middle Ages.

The rodents

Several species from the order of rodents (Rodentia) have either passed completely into the ranks of domestic animals or are today at least used in the same ways as the older domestic animals from this group in a preliminary stage to domestication. In their case, the possibility of applying the preliminary definition of domestic animal given in Chapter 1 becomes more and more vague in contrast to the fairly clear situation with respect to large mammals. The oldest domestic rodent in the Old World is the fancy mouse, derived from wild house mice (*Mus* species, *musculus* group), which played a part in Greek Bronze Age cults. Aristotle mentioned the albino form, the universally known white mouse (Fig. 3.45). The Romans used this type for fortune-telling, and it was also in use in ancient China. Spotted mice are mentioned in the Chinese Bronze Age. Fancy mice reached Europe from Japan in the middle of the nineteenth century. Along with this species, the laboratory rat belongs in the mouse family (Muridae). The first mention of albino laboratory rats comes from 1856. The black and hooded mutants also came up in these years in England. Their origin from the brown rat (*Rattus norvegicus*) can be traced to some extent. From the turn of the eighteenth to the nineteenth century, the organization of rat-baiting by terriers was popular for some decades in France and England, and then also in America. In the manner of a 'blood sport', 100 or 200 freshly captured brown rats were released into an arena and the dog had to kill them all in as short a time as possible. According to the records, the albinos, which, though rare, were nevertheless repeatedly found amongst the huge numbers of rats needed, were picked out. Instead of being put into the arena they were put on show, for which purpose they were then also bred. So in this case the display value led to domestication (Fig. 3.46).

Another colour variant of a wild species, the Syrian hamster (*Mesocricetus auratus*) from Southeast Europe and Asia Minor, from the family

Fig. 3.45. Albino house mouse.

Fig. 3.46. Albino laboratory rat.

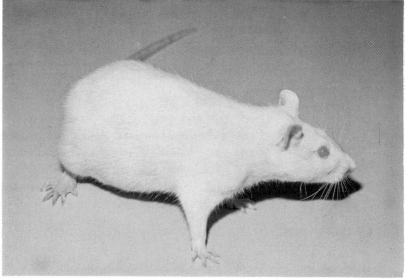

Cricetidae, initiated the domestication of an equally widespread rodent, the golden hamster. All the golden hamsters found today in homes and laboratories come from a female with 12 young caught near Aleppo in Syria in 1930 that were used to build up a stock in the Zoological Institute of the University of Jerusalem. The first animals from this breed came to England in 1931 and to America in 1938 from where they reached Germany in 1945 (Fig. 3.47).

In the New World, the domestic guinea pig was bred as an important source of meat from the wild guinea pigs (*Cavia aperea*; suborder Caviomorpha) in the tableland of Peru in pre-Inca times (Fig. 3.48). Two other species, originally South American and important today as fur ranch animals, the chinchilla (*Chinchilla laniger*) and the coypu or nutria (*Myocastor coypus*), belong to the same suborder. In addition to the already 'classical' laboratory animals, mouse, rat, golden hamster and guinea pig, some further species have recently been taken up for breeding as laboratory animals but this can only be regarded as a first stage of domestication at the most. They come from the family Cricetidae and are

Fig. 3.47. Golden hamster.

Fig. 3.48. Domestic guinea pigs with a diversity of colouration.

two species of gerbils (*Meriones unguiculatus* and *M. shawi*), the cotton rat (*Sigmodon hispidus*) (Figs. 7.3 and 7.4 (pp. 118–19)) and the Chinese dwarf hamster (*Cricetulus griseus*),

As has become clear from this review of the origin of the domestic mammals, the date of domestication cannot be ascertained at all for some species such as the reindeer, yak, Bali cattle and mithan, while for others such as the llama and alpaca the range of uncertainty spans several thousand years. Attempts to date domestication are subject to a high margin of error for the majority of species. The problem of distinguishing with certainty between the remains of wild and of primitive domestic animals in the archaeozoological records from the central areas of primary domestication has been mentioned several times. So it is possible that finds interpreted as 'domestic animals' come from wild animals, perhaps in the stage of being farmed as game and vice versa. Furthermore, the chance nature of the finds plays a large role in the determination of the initial stages of domestication so that, even where it is possible to identify remains as coming certainly from a domestic animal, it cannot be ruled out that farming the species in question began considerably earlier and that further finds could push the date back much further than is estimable. Consequently, any picture of the distribution of mammal domestication over the millennia can only be drawn up with reservations on the basis of what is currently known (Fig. 3.49). Nevertheless, such a picture seems to permit one fundamental proposition, namely that there was no period in pre-history in which domestication clearly took place at much increased frequency on a world-wide basis. The distribution of the domestication dates for all those animals for which

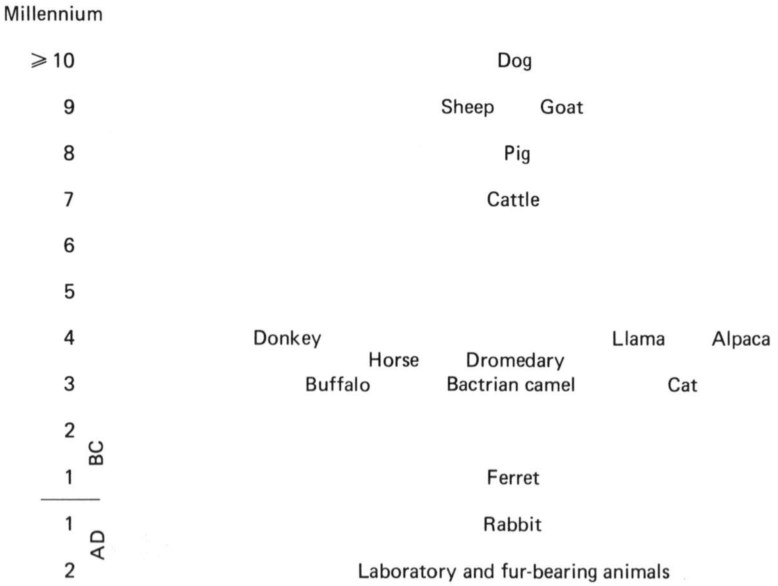

Fig. 3.49. Scatter of the dates of origin of various domestic mammals. A fairly wide span of uncertainty must be allowed for in each case. Species with domestication dates which are still completely unknown have not been included in this diagram.

Millennium					
⩾ 10			Dog		
9		Sheep	Goat		
8			Pig		
7			Cattle		
6					
5					
4	Donkey			Llama	Alpaca
		Horse	Dromedary		
3		Buffalo	Bactrian camel	Cat	
2 BC					
1			Ferret		
1 AD			Rabbit		
2			Laboratory and fur-bearing animals		

dating seems to be possible, at least to approximately one or two millennia, dispersed over time. Wherever a certain increased frequency might be read into the picture, it disappears when the variability in regions and civilizations are taken into account (Fig. 3.50). Thus, the first dates for the sheep, goat and pig indicate only a possible proximity in time, as they come from the Near East for the first two species but from New Guinea for the last. The same holds for a certain clumping for the period from the fourth to the third millennium BC, in which the domestic animals first known with certainty come from regions as far apart as East Europe, Soviet Central Asia, South Asia, the Arabian Peninsula, North Africa and South America. So chance does indeed seem to have played a leading role in domestication.

While success in gaining a new species of domestic animal is to be regarded as an aspect of this element of chance, this need not be true for the production of new useful animals. The idea of keeping animals may indeed have expanded constantly from the Neolithic onward and it seems quite conceivable that in later times and civilizations it culminated in numerous attempts to keep and breed animals. The results, however, the abiding stock of domestic animals, display no clear relationship to any such possible increased endeavour. Only a fraction of the animals kept in captivity eventually became domesticated, as was pointed out at the beginning of this chapter. It was obviously a question of the chance combination of an attempt or desire to keep an animal species and the special suitability for domestication of the species which happened to be available, i.e. that the right animal was at the right place at the right time. Several fundamental prerequisites are to be understood as the suitability of a wild animal for domestication. The relevant animal had to be easy to

Fig. 3.50. Areas of domestication of the larger domestic mammals (excluding rodents). There is still considerable uncertainty on the area of primary domestication for some species.

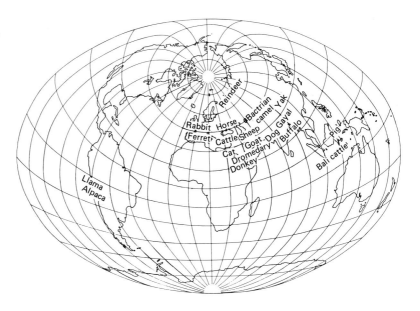

tame and master; it must not have been so aggressive towards other members of its own species as to be impossible to keep constantly in groups, so that breeding under simple circumstances could be carried out. Finally, it had to reproduce successfully in such conditions of captivity. These minimum requirements all relate to the broad field of behaviour. The suitability for domestication had to manifest itself in behavioural features because becoming domesticated presupposes behavioural adaptation to the new environmental conditions – constant contact with humans in contrast to life in the wild. Consequently, a fundamental knowledge of this change in behaviour is necessary to the further understanding of the domestication process and will be treated in the next chapter.

Synopsis

Becoming domesticated presupposes that a wild animal is suitable for the process and that people have an interest, for whatever reason, in keeping the animal. An endeavour to keep numerous species of animals in captivity, such as is demonstrable from the third millennium BC onward at the latest, would not alone be sufficient to increase the stock of domestic animal species if the first prerequisite were not fulfilled. No clearly distinct period of increased domestications can be detected in the dates of origin of domestic animals, the spatial distribution of which is scattered widely over the millennia. However, these dates can only be determined with large ranges of uncertainty, in most cases due to the difficulties in distinguishing conclusively the wild from the primitive domestic animal remains in the archaeozoological data from the areas of primary domestication, areas moreover which have not yet been clarified for all species. The domestication dates for some species are still completely uncertain.

4 | Changes in behaviour

To detect similarities and differences in the behaviour of a domestic animal and that of the corresponding wild form, studies carried out under comparable conditions are necessary. So far such studies, in which domestic and wild animals living in comparable conditions of captivity, or free-living wild animals and feral ones, have been observed and compared, have been carried out only for a minority of the possible pairs of comparable animals. Unfortunately, many quite far-reaching hypotheses on the changes in behaviour resulting from the transition from the wild to the domestic state are based on comparisons of domestic animals, completely under human control, with the behaviour of free-living wild animals. It is easy to see that these theories can scarcely be reliable if the goal is to find out about the behavioural differences due to domestication rather than about those due solely to the conditions of keeping.

The classification of behavioural deviations

Using examples taken mostly from domestic poultry, Konrad Lorenz classified behavioural deviations observed in domestic animals into several groups: hypertrophied and atrophied instinctual activities; expansion of innate releasing mechanisms; dissociation of correlated modes of behaviour; and persistence of juvenile characteristics. He conjectured that the causes lay in quantitative alterations in the production of internal stimuli, i.e. in the production of central nervous energy. This theory may be subjected to critical examination using, for example, peculiarities in the behaviour of the domestic cat as are suggested by Paul Leyhausen to be attributed to domestication.

Atrophy (the reduction or disappearance of any one innate activity) is found amongst domestic cats insofar as some catch prey such as mice, though they do not kill it, at least not intentionally. This seems to be largely dependent on experience, i.e. on what the cat learns when young, and is consequently modifiable, which is to say this behaviour is alterable and is independent of changes in heredity. Hypertrophies, overexaggeration of individual activities, can also be found in the prey-catching behaviour of domestic cats, for example when omitting killing is accompanied by an increase in the catching activity, which leads to very 'playful' cats. When they have been reared under corresponding conditions, however, some captive wild cats exhibit similar activity; this therefore also can be seen, at least in part, as a modification that tells us little about the real difference between a wild and a domestic animal. Prey-catching behaviour also supplies instances of dissociation, i.e. the breakdown of an activity chain and, when seen from the viewpoint of the original function, the senseless performance of single activities comprising the chain. Bringing in prey without consuming it is a common activity in the course of raising young. Some domestic cats bring in dead or living prey completely outside the context of caring for young, when they catch prey, drag it back into the house or lay it down in front of the human companion without eating it themselves. All of these cases may suggest some transfer of caring for kittens to humans or subconscious training by the owner, with administration of rewards and punishments interfering in the chain of activities involved in the completely normal bringing of prey to a customary feeding place.

When judging such alterations in behaviour, the large part played by different learning conditions in the human environment of a domestic animal and the free environment of a wild animal should never be overlooked. A persistence of juvenile behaviour occurs in domestic cats in several respects. As a rule, social compatibility ceases at the age of six months in European wild cats, while it can endure into old age in domestic cats, especially between siblings from one litter. A certain continuing compatibility, however, has also been found in African wild cats (the probable ancestors of the domestic cat, see Chapter 3) in captivity and even occurred in individual cases of European wild cats kept in confined conditions. Purring seems to be significant in wild animals only during the development of the young in the communication between mother and kittens but is generally known in the domestic cat, above all in communicating with humans. In similar conditions of captivity, however, it occurs in the same way not only in adult African wild cats but also in many other cat species from several genera. This indicates, at least in part, that the cause is modification due to environmental changes. Of course, the totally different environmental conditions in which domestic and wild animals normally live involve a whole series of obvious differences in behaviour which should not be confused with innate differences in the organization of behaviour.

As in this example where the behaviour of the domestic cat was related to that of the European wild cat, from which, however, it does not descend primarily, the behaviour of other domestic animals has often been compared with that of geographically separate forms of the wild species from which they did not originate or which were at most involved only in the secondary domestication of the breed. Since behavioural traits, like external features, are subject to change between populations, differences between the organization of the behaviour of different geographically separate populations cannot be ruled out from the start. This limits the strict comparability of the behaviour of domestic and wild animals not only to the same observational conditions but also to individuals from the populations of the wild species from which the domestic animal actually derived. Up to now, there have been very few useful attempts at comparison that meet these conditions. So, with this reservation in mind and in order to obtain a broader basis for verifying generalizations, the following discussion will still include comparisons with other geographically separate forms of the relevant wild species.

The wolf/dog pair

The behaviour of the wolf/dog pair has been studied by Erik Zimen with wolves and dogs kept in kennels. However, he used northern wolves, which are not close to the origin of the domestic dog (cf. Chapter 3) and poodles, as representatives of the dog in general; that is, he proceeded from an improved breed. In comparison with the wolf, the poodle exhibited great changes in behaviour due partly to less agility and the lower intensity with which many activities were carried out. Many modes of behaviour of the adult poodle are comparable to those of young wolves, thus corresponding to the pattern of persistence of juvenile characteristics. The distribution of daily activity in the adult wolf has two distinct peaks, one in the morning and one in the evening, whereas the young wolf switches more rapidly between phases of activity and rest, and the morning and evening peaks are less distinct. So the geophysical time synchronizer provided by the change in light conditions for the distribution of activity over 24 hours affects adult wolves more strongly that it does young ones. In this respect, the poodle behaves like the young wolf; that is, its activity organization is also less affected by this general time synchronizer. Due to their more rapid switching from activity to rest, both the poodle and the young wolf are easier to activate at any time than adult wolves. Thus, as for other behavioural characteristics, one can speak of a general reduction in the intensity of activity in poodles as compared with wolves.

Furthermore, the social organization of the dog seems to be less differentiated than that of the northern wolf. Poodles behave more like young wolves, the distance individuals keep from each other being shorter and the active contacts more frequent. On the whole, the poodle lacks the strict control of the wolf in its movements, the intensity of many forms of expression, and the complexity of the social order in the wolf pack. The poodle seems to be less highly developed in most traits, weaker in its activity and reaction, though vocalization is an exception: barking plays a large role in the poodle's expressive behaviour but not in that of the wolf. The movements of the facial musculature and of the tail, important as communication symbols, seem to be less differentiated and clear in the dog as compared with the wolf. The vocalization seems to be simplified: Wolf Herre has summarized these differences as reduced informational content. As he pertinently formulated it: domestic dogs do not have as much to say to each other as do the members of their ancestral species. Michael W. Fox pointed out that the level of general activity and of exploratory behaviour is very high in the wolf, whereas in these traits the variability, particularly, in the dog is very wide.

Comparisons between different breeds of dog, such as those carried out by Lothar Schilling and Michael Schmidt from the author's research group, provides further information on the change in behaviour from wolf to dog. The breeds used for the comparative study were: dingoes, as examples of a very primitive dog coming from feral populations; Siberian huskies and Greenland sled dogs, as examples of northern primitive dogs with a higher intermixture of northern wolves; dalmation–setter crossbreeds, as examples of improved breeds; and alsatian backcrosses of a northern wolf–alsatian mongrel. It turned out that the behavioural extremes between the northern wolf and the poodle in Erik Zimen's studies are partly bridged by the great developmental span between the primitive dog and the improved-breed dog. With respect to the organization of activity, the ease of activiation and the barking, the only one of the above forms that can on the whole be placed alongside the poodle is the crossbred hound. The activity organization of the dingoes as observed by Lothar Schilling places them closer to the wolf. Although they switch between activity and rest more frequently, main and subsidiary activity maxima can be distinguished and the transition from resting to activity is rather abrupt. The resting phases are shorter in the wolf; higher movement intensity in the activity phases leads to a much higher level of activity in the wolf. Disciplined movements contrast with a much lower differentiation in the social interactions of the dingo compared with the wolf. Aggressive behaviour comes to the fore in social life; the social compatibility seems to be much lower in comparison with the northern wolf. This means that, when free, dingoes live alone or in pairs and adult animals only exceptionally form a pack of several animals. In this respect, the dingo's behaviour is more like that of the coyote and especially that of the southern wolf, the main ancestor of the dog.

With respect to the distribution of their daytime activity the northern sled dogs also come very close to the wolf, although in the given observational conditions an environmental influence obviously contributed to the formation of marked morning and evening peaks. The insulating effect of their long coats causes a drop in activity in the warm hours of the day, the distribution of activity being modified by the behaviour involved in temperature regulation.

The wild cat/domestic cat pair

Erich Zimmermann studied wild cat and domestic cat under comparable conditions, mainly with a view to litter care behaviour and the behavioural development of the young. He found a reduction in the significance of the seasons as a time synchronizer for the distribution of births. Births in European wild cats are very closely coupled to the seasons. The curve for birth frequencies over the course of the year in European short-haired domestic cats and Persians flattens so that the maxima and minima are only about half as marked, and this becomes still more uniform in the case of Siamese cats. Behavioural deficiencies are to be observed mainly in the Persian but also in the Siamese cat with respect to freeing the newborn from the afterbirth, severing the umbilical cord and eating the placenta, lying down with the young, transporting them and defending the nest. It seems that the relevant stimuli have reduced significance for these cats. The behaviour necessary for unproblematic rearing of the young either does not take effect at all or only in a weakened form.

Similarly, aggressive feeding in young and adult cats becomes weaker in the following sequence: European wild cat →European short-haired domestic cat → Siamese cat → Persian cat. The ontogenic development of the various ways of moving is delayed in improved breeds. Social playing persists even into adulthood and shows that social compatibility somewhat exceeds that found in the African wild cat but is a great deal higher than that of the European wild cat. Heidrun Müller from the author's research group carried out observations, which were mainly quantitative, of the behaviour of European wild cats, African wild cats and all important breeds of domestic cat under approximately comparable conditions. With respect to social relationships, she found that the European wild cat can be characterized as a 'distance' type and the African wild cat and the breeds coming mainly from Southeast Asian domestic cat populations (Siamese, Oriental shorthair) as 'contact' types. The European short-haired cats, probably crossbred with European wild cats in the course of their racial history take up an intermediate position.

All breeds of domestic cat distribute their daily activity and rest phases much more evenly on the average than the two wild forms, to which the European short-haired cats come the closest. The activity and consequently at least partly the social behaviour of the very long-haired Persian cat seems to be much more attenuated due to the necessities of temperature regulation.

The ferret/polecat pair

Quieter behaviour is described for the ferret in comparison to tame polecats. They are said to be slower in all their movements and not so easy to frighten. Above all they make hardly any use of their anal scent gland so that they are much more pleasant to keep. The organization of their daytime activity seems to be more balanced than that of the polecat. As in the comparison between wolf and poodle, there is a decrease in the intensity of activity of the domestic animal.

Comparisons of wild and domestic equids

Although there are several field studies of domestic horses either returned to the wild state or living mostly untended, there can be nothing comparable for their wild horse ancestor. This is because this direct progenitor is extinct. It would now be possible to carry out studies only on the Przewalski horse, which is threatened with extinction, if not yet quite extinct, in the wild state. Even observations made under comparable conditions and with comparable methods on Przewalski horses and domestic horses in zoos are available in only the initial stages and no clear behavioural differences have been detected. Studies of feral asses by Susan L. Woodward show that these animals form small, highly unstable groups, older males tending to be solitary. The territorial system of male wild asses seems to break down in these domestic relatives, and the restricted breeding season of the wild asses may have been lost.

Zoo studies on a uniform basis are still in their infancy. It seems that the daily activity distribution of the domestic donkey is characterized by a more rapid succession of active and resting phases than that of the wild ass, but reliable statements are not yet possible. From quantitative observations carried out by Eckehard Eich and Elisabeth Reichert from the author's research group, it seems that the two domestic equids horse and donkey can always be activated more easily than the Burchell zebra, another wild species of the horse family. An index calculated from the duration of activity and rest phases and the frequency of phase alternation is about twice as high for the normal-coloured zebra as for the horse and donkey. Zebras have longer activity phases and so less rapid alternation between high and low motor activity.

The wild pig/domestic pig pair

Manfred Röhrs and Dieter Kruska have recorded a series of behavioural differences between domestic and wild pigs but gave no details on the comparability of the observations on which their findings were based.

The constant attention and vigilance of the wild pig is said to be largely missing in the domestic pig. The social structures of the wild pig are relaxed in the domestic pig and its motility, i.e. the sum of its spontaneous locomotor activity, is lower. Both the flight distance and the aggression of the domestic pig are also lower, as are its nest-building activity and defence of the young. On the other hand, the sexual behaviour is more intense and the number of young higher. Other observations indicate less clear division of daily activity in the domestic pig. Studies by Konrad Wolpert from the author's research group confirm these observations on the whole. Wild pigs synchronize their behaviour within the group more than domestic pigs and make social contacts more often; aggression and motor activity are much higher and the activity phases longer.

The guanaco/llama pair

Hilde Pilters has undertaken very detailed but mainly qualitative behavioural studies of the wild–domestic animal pair guanaco/llama under comparable zoo conditions. They revealed clear deficiencies in the mating behaviour of the llama. The movements made by the male guanaco in the foreplay to mating – snapping, circling around the female and pressing down her neck – were mostly lacking in the male llama. By contrast, male llamas chase, with great perseverance, even females not ready to mate, and copulate with them if their defence is weak, which does not happen in the guanaco. Female guanacos not ready to mate attack the male but female llamas do not and their readiness to mate lasts longer in each oestrus. Male llamas also reach sexual maturity earlier than guanacos. So in the sexual behaviour of the llama a lower intensity is coupled with a generally stronger sexuality.

Comparisons of wild and domestic reindeer and cattle

Wild reindeer do not herd uniformly throughout the whole year and their groups are smaller in summer than those of domestic animals. Comparisons between domestic cattle and their wild counterparts are no longer possible for the domestic cattle/aurochs pair, and have not yet been carried out with uniform methods and under uniform observational conditions for the other wild species. Small groups of five to eight animals are reported to be typical for obviously wild water buffalo in Sri Lanka while feral buffalo form much larger herds. Further comparative data are available for the small domestic ruminants sheep and goat.

The moufflon/domestic sheep pair

Field observations of released moufflon and feral sheep on Hawaii have shown that, under comparable conditions, moufflon live in small bands but domestic sheep live in large flocks, so that the latter cause much more damage to vegetation than the former. In the author's research group, Wolfgang Pees carried out quantitative studies on moufflon and several breeds of domestic sheep, namely Soay sheep, Hungarian Zackel sheep, German heath sheep and Texel sheep kept in runs. His results disclose a gradual change in behaviour from the 'wild' species, the moufflon, via the primitive domestic sheep, the Soay, to the improved woolly sheep breeds. Motility is highest in the moufflon with about 22% rapid locomotion within the whole period of observation, followed by the Soay sheep with about 18%, the German heath sheep and Hungarian Zackel sheep both with about 15%, and finally the Texel sheep with only 7%. Moufflon and Soay sheep lie down longer than they stand or graze, but the improved woolly sheep breeds stand longer without moving rapidly. The flight distance of the moufflon is the longest, being between about 25 and 30 m. In the Soay sheep it is distinctly less, about 12 m, while the German heath sheep and Texel sheep take avoidance reactions only at about 4 to 7 m. Even after several days of becoming accustomed to the observer in the run, the avoidance distance hardly decreased in the improved breeds, while it did in the Soay sheep and moufflon.

Moufflon and Soay sheep exhibit a high degree of social integration; apart from the occasional separating off of the rams, the members of a group usually rest very close together (Fig. 4.1). Once any particular sheep has spontaneously stood up, the other members of the group quickly end their rest one after the other. When they flee they all stick close together. By contrast, German heath sheep and Texel sheep rest in

Fig. 4.1. Gregariously living ancestors of domestic animals as a rule are characterized by a closer group cohesiveness than the domestic animals descended from them. The activities of individual members of a group are more closely coordinated with those of the others, and the groups are mostly smaller than domestic animal groups. An example is provided by the comparative observation of moufflon and domestic sheep. (A band of female moufflon and lambs acting as a close community.)

loose groups from which individuals may separate themselves for quite a long way. There is much less co-ordination in their lying and moving phases (Fig. 4.2). In sum, the group cohesion is less strong than in the wild sheep and the primitive domestic breeds.

The wild goat /domestic goat pair

Different herd sizes for bezoar and feral goats have been reported from the Aegean, where male groups of domestic goats comprise mostly 15 to 20 animals, with flocks of 30 to 40 or more also occurring. The mean herd size of the bezoar goat is only about four to ten animals, which agrees quite well with the numbers of three to seven reported from the part of their distribution area which lies in the Soviet Union. Group sizes of three to five are given for the markhor, and five to ten for the ibex. Only in the winter oestrus do larger herds of ibex gather together. Wild goats live mostly in rather small groups and feral goats in larger herds.

Under zoo conditions, Ursula Mayer from the author's research group carried out quantitative behavioural studies on African dwarf goats (as representative of the domestic form), Alpine ibexes and markhor. It was only possible to make pilot observations of bezoar goats. She was able to show that the peaks in daily activity are less distinct in the domestic animals, their activity being fairly uniformly distributed over the day in rapid sequences of activity and rest periods. This uniformity recurs in the course of the year where the clear division of seasonal behaviour due to oestrus does not take place, as it does in the wild goats. Oestrus is more widely distributed and affects the behavioural organization of the group less markedly in domestic animals. Taking into account the seasonal and daily fluctuations, the domestic goat's motility tends to be lower and the

Fig. 4.2. Social bonds are as a rule looser amongst domestic animals than between their wild ancestors. Behaviour seems to be less coordinated, the cohesion less close in the generally larger groups. (Mallorca: domestic sheep grazing loosely separated from each other in a rubble-walled enclosure of a type used probably since the Neolithic.)

intensity of movement, the liveliness as expressed in the proportion of rapid movements in the total movement, is reduced. The billygoats fight less often and less intensely, the mother–kid bonds are looser than in the wild species ibex and markhor. Overall group cohesion is lower, the activities of individual domestic goats being less related to each other than are those of the wild forms under study; social bonds are therefore weakened. The same acoustic stimuli elicit less strong reactions in domestic than in wild goats. In such situations, ibexes and markhors emit warning calls and their groups draw close together to take flight. Finally, the flight or avoidance distance of the domestic goat is shorter than that of the wild animals.

The wild rabbit/domestic rabbit pair

The comparison between wild and domestic rabbit has been studied in detail mainly by Richard Kraft. He detected less dexterity and agility in the domestic than in the wild rabbit. The domestic rabbit's readiness to flee is considerably diminished, involving certain deficiencies in protective and defensive behaviour (Fig. 4.3). During the daytime rest periods, the domestic rabbit remains outside its burrow whereas the wild rabbits disappear into theirs. Aggression, sexual and marking behaviour, however, are stronger in the domestic rabbit. The activity of the domestic rabbit is distributed more evenly over the whole day than that of the wild rabbit, while the differences between day and night activity are weaker.

The wild cavy/domestic guinea pig pair

The behaviour of wild cavies and domestic guinea pigs has been studied comparatively in captive groups by Adelheid Stuhnke. The social

Fig. 4.3. Feral rabbit, found resting outside its burrow under a bush during the day, which had not turned to take flight even when approached to within a few steps. This behaviour seems to be typical for domestic rabbits, whereas wild rabbits usually disappear into their burrows to rest during the day and possess a much higher readiness to flee. The animal in this photo exemplifies that in certain environments, even the domestic animal's spotting can provide excellent camouflage. The yellow and black patched coat blends excellently into the play of light and shade on the ground.

tolerance of the wild species is clearly less than that of the domestic one: there are more agonistic behaviour patterns towards conspecifics in the former, whilst the latter act more indifferently towards conspecifics in neighbouring cages. Wild cavies react more, with higher intensities and in a more differentiated manner to external disturbances than do guinea pigs.

Comparisons of wild and domestic rats and mice

Studies of brown rats and laboratory rats undertaken by Curt P. Richter show looser social ties in the domestic form, which huddle together much less closely than brown rats. The reactions of the laboratory rat when alarmed are weaker; when held tightly they do not panic. Their aggressiveness is reduced and so is their motility, even when hungry. Similar findings resulted from comparisons between wild house mice and laboratory mice. Here as well, the latter exhibit less readiness to flee and less tendency to panic reactions. The time course of behavioural organization was compared quantitatively in the author's research group, mainly by Bernd Rosenbaum. The distribution of activity of laboratory mice was more uniform over the whole 24 hours and there is less difference between diurnal and nocturnal activity. White mice living under the same conditions as wild ones seem to enter traps more readily.

Conclusions

The behavioural differences between wild and domestic animals summarized here have a number of common aspects implying the following generalizations. Compared with their wild ancestors, domestic animals:

(1) exhibit less intensity ranging down to a complete lack of various patterns of behaviour including reductions in care of the young and in motility, i.e. there are attenuations in behaviour;

(2) are less ready to take flight and have weaker alarm reactions;

(3) have less overall activity and distribute their activities more uniformly over the course of the day, i.e. the influence of the geophysical time synchronizer, the alternation between day and night, is weaker and consequently they are more easily activated at any time – likewise seasonality is weaker;

(4) have looser social bonds combined with a decrease in social complexity and social differentiation and often an increase in social compatibility;

(5) have intensified sexual behaviour, and perhaps sometimes intensified intraspecific aggressiveness.

These points can be subsumed under even more general categories. Points (1) to (4) are various expressions of a general attenuation in the behaviour of domestic animals; put broadly, domestic animals live less intensely than wild ones. Points (2) to (4) involve a general reduction in the significance of factors from the inanimate and animate environment including the social environment for domestic animals. This situation can be described with the term environmental appreciation coined as *Merkwelt* by Jakob J. von Uexküll and revived here. Von Uexküll originally meant by this term the totality of the components and features of the environment of a living creature as it perceives them via its sensory organs, i.e. as it notices them. This should be enlarged to include how the living creature evaluates what it perceives by processing it using the constituents of its memory. Expressed in this way the environmental appreciation of domestic animals looks to be characteristically reduced in comparison with that of their wild progenitors. The weaker reactivity of the domestic animal in turn leads back to the first inference concerning lower intensity of life. At first sight, point (5), the increase in the intensity of sexual behaviour, seemingly contradicts the foregoing.

Synopsis

The behaviour of domestic animals seems to be weaker and less determined by environmental factors in contrast to that of wild animals. The environmental appreciation of the domestic animal is innately reduced when compared to that of wild animals. This is expressed in a lower intensity or even disappearance of particular patterns of behaviour, in a drop in motor activity, in more uniformity in the temporal organization of activity, in a loosening of social bonds and in a breakdown of social differentiation. A singular intensification of sexual activity contrasts with the general attenuation of other behaviour.

5 | Stress

Stress is to be understood here in a strictly limited sense as the state of general activation of an organism due to stimuli. Such stimuli (or stressors) can be of a physiological as well as a psychogenic nature. Physiological stress results from factors, such as cold, heat, lack of food, or oxygen deficiency, that directly affect organic activity. Psychic stress is based mainly on signals from conspecifics, so a main component is the psychosocial stress caused by the social environment. The dependence of many organic functions on the population density of a species, i.e. the number of individuals per unit of space, clearly documents this kind of stress.

Many studies in this direction have been carried out, particularly on rodents (Fig. 5.1), but confirmation of the results obtained has also come from other orders of mammals, including the primates, so that the fundamental precepts can be generalized. A rise in population density is accompanied by a retardation of the animal's growth leading to a smaller final size. The body weight decreases, the beginning of sexual maturity is delayed and even entirely suppressed in extreme cases, the oestrous cycles are lengthened, the number of young drops as does their life expectancy. At the level of single organs, the adrenal cortex becomes

Fig. 5.1. The influence of psychosocial stress, as it arises from impinging signals in groups of conspecifics, has been particularly well studied in rodents. In the European common hamsters (*Cricetus cricetus*) in this photo, raising and keeping a whole litter together in a very confined cage led to considerable inhibition of the bodily development and failure to reproduce. Finally, several individuals were killed one after the other by their conspecifics.

larger, the sexual organs are reduced and the formation of sperm cells in the male is retarded. Ovulation and the reception of fertilized eggs in the uterine lining diminishes in the female. The prenatal death rate in the womb rises. At the biochemical level, one effect of the higher activity of the adrenal gland is a rise in the level of glucocorticoids in the blood, i.e. the hormones secreted by the cortex of the adrenal gland that influence sugar metabolism. This involves an inhibition of the immune system, so that resistance to infectious diseases is reduced and mortality increases.

Central sites of stress events are the hypothalamus and the pituitary glands. The stimuli, signals or information received from the environment and transformed into excitation affect the hypothalamus via the thalamus of the diencephalon, depending on how they are processed for significance in the mammalian neocortex. The hypothalamus, a small area at the base of the brain, above the pituitary gland, causes excitation in the sympathetic nervous system that activates the body and alerts it so that it can react unhesitatingly with fight or flight in cases of acute danger. This is the acute stress component (fight and flight syndrome). However, the hypothalamus controls the production of hormones in the pituitary gland and these hormones regulate other hormone glands in the body. This is how the chronic stress components expressed in the effects of populations density arise (Fig. 5.2). Following a stimulus load, the pituitary reacts by increasing the secretion of adrenocorticotrophic hormone (ACTH), which activates the adrenal cortex. The increase in corticosteroid hormones produced there causes the higher susceptibility to infection mentioned above, loss of weight and retardation of growth, increases the responsiveness of the adrenal cortex to ACTH and intervenes in the activation of the thyroid, which is controlled by the pituitary

Fig. 5.2. Simplified diagram of the central stress axis from the input to the sensory organs (receptors) via information processing in the brain to the activation of the adrenal cortex and increased secretion of corticosteroid hormones (general adaptation syndrome). Some subsidiary and feedback pathways decisive for the overall state of the organism are included. The pathway to the sympathetic nervous system (not shown in detail) is that of acute or short-term stress, which leads to direct activation of the body upon high momentary stimulus load (fight or flight syndrome). TSH, thyroid-stimulating hormone; ACTH, adrenocorticotrophic hormone; FSH, follicle-stimulating hormone; LH, luteinizing hormone.

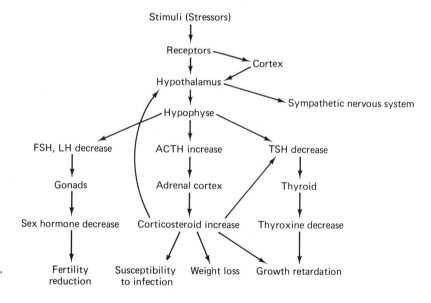

via the secretion of thyroid-stimulating hormone, making possible further influence on growth. At the same time, the pituitary lowers the production of the hormones stimulating the activity of the gonads, i.e. follicle-stimulating hormone and luteinizing hormone. This reduces the activity of the gonads, resulting in retarded sexual maturation in juveniles and diminished fertility in both sexes.

Since stress comes about through the influence of stimulus load, its intensity must be related to the quantity of such stimuli. This is directly demonstrable from the dependence of psychosocial stress on population density. An increase in population density means an increase in the number of conspecifics constantly around an animal or an increase in the frequency of encounters with conspecifics. Signals from other animals that constitute loading stimuli for the one in question increase in number per unit time as the population density increases. So stress can be defined as a mathematical function of the stimuli impinging on an individual and the information it acquires by processing them. At this point, we can see a connection with the change in behaviour between wild and domestic animals described in the preceding chapter. The comparison with wild animals revealed that the behaviour of domestic animals is determined through the lower significance of environmental factors, of stimuli of a geophysical, interspecific and social, intraspecific nature, in brief through a decline of environmental appreciation. Consequently, less stress is to be expected in domestic than in wild animals in objectively equivalent environmental conditions.

Many different observations confirm this prediction. Situations that involve enormous stimulus load and so may easily overtax the organism of a wild animal occur, for example, when it is caught. In certain circumstances, holding a wild rat firmly can cause a violent panic reaction leading to death from stress within a few minutes. Wild mice can also die of shock after being hunted incessantly and finally caught. Such over-reactions seldom occur in laboratory rats and mice. The stress organ essentially responsible for activation, the adrenal gland, is generally smaller in domestic animals than in wild ones. This can be interpreted as a consequence of chronic high stress in wild animals and of diminished stress in domestic ones. Because of the decline in their environmental appreciation, domestic animals respond less in conditions that elicit strong stress reactions in wild animals. To reach the same level of psychosocial stress, a smaller number of social partners – a smaller group – should be sufficient for the wild animal as compared with the domestic one. This agrees with observations of larger herds being formed by domestic animals such as sheep, goats, water buffalo or reindeer (Chapter 4) who are scarcely aware of food competition and the aggression it provokes. The herds formed by the relevant wild species are typically much smaller. At the same time, the apparent contradiction between a general attenuation of behaviour and an intensification of sexual activity in the domestic animal is easily reconciled with the stress concept. Less stress means in the end a higher production of sex hormones, which influence not only the reproductive performance itself but also male aggression. It must be expected that domestic animals subject to less stress thanks to their reduced environmental appreciation will be more active and successful in their reproductive biology than will wild animals subject to increased stress. Naturally, all these comparisons are valid only under comparable environmental conditions. When the conditions in which it is kept are unfavourable, a domestic animal will be subject to as much stress, in an absolute sense, as a wild animal in its natural habitat. Following this line of argument, a concept of increased psychogenic tolerance can be formulated:

Compared with their wild ancestors, domestic animals are characterized by an innately reduced environmental appreciation and so by increased psychogenic tolerance in comparable conditions.

Since life in a group plays a central role in the occurrence of psycho-social stress, this implies a first general principle of domestication with reference to the special suitability or potentiality of a species for domestication:

The easier it is to breed wild animals successfully when they are kept in groups in confined conditions of captivity, the better suited they seem to be for domestication.

According to this principle of domestication, data collected of a species kept in zoos, concerning the regularity of progeny, the mortality of the young and the beginning of sexual maturity, acquire a key role. Clues can also be gleaned from field observations of social behaviour, which has a bearing on the success of keeping a species in captivity. Thus, it is found that most, though not all, wild ancestors of domestic mammals live gregariously.

Wolves are known to live in packs, in which complex mechanisms to control and stabilize social life are developed; they possess a high capacity for socialization. This, however, is only valid for the highly evolved northern wolves. There are fewer observational data on this subject for southern wolves, the likely ancestors of the dog. It seems, however, that the drive for socialization beyond the core family is much lower in the wolves of the Arabian Peninsula and India than in northern wolves, although there are faint indications that the former tolerate crowding in captivity better. They pass their lives more frequently in pair bonds or as solitary animals, as do the coyote and the dingo, and as may well be the case for primitive dogs living free in other regions. Wolves turn out to be quite easy to keep in captivity where also they have no problems in reproducing.

In contrast to the wolf, the wild cat obviously has little inclination for any direct socialization outside oestrus. As far as is known, this also seems to hold in general for the African wild cat. On the other hand, African wild cats do tolerate being kept in larger family groups in captivity, without any detectable disturbances in their reproductive behaviour. The fur animals, fox and mink, and the ferret, are also descended from more or less solitary wild ancestors. They are special cases insofar as these species are not usually kept together in large groups in small cages or runs so that at least psychosocial stress plays a subordinate role. Although the red fox may be kept in groups, this frequently gives rise to problems when they breed. The ferret, a true domesticated species, tolerates being kept in groups.

Wild horses and asses are moderately gregarious animals that also tolerate living in groups and reproduce in captivity. Wild pigs are quite markedly gregarious: their fertility does not suffer in any noticeable way when they are kept in enclosures. Wild camel species also live in social groups. Little is yet known about any possible problems in keeping the wild two-humped camel in captivity. Caring for guanacos in groups in zoos is straighforward, whereas with vicunas there are breeding problems that are attributed to the excitable nature of the males.

Wild tundra reindeer periodically form the largest bands of all deer species. That it is not easy to keep them in zoos much farther south than their distribution area is probably due to unsuitable climatic conditions and diet, particularly since similar problems occur with domestic reindeer. Pure physiological stress probably plays a primary role, without any significant psychosocial component. Some wild cattle also live in large herds. The gaur used to be considered to be difficult to keep in captivity but this is apparently not entirely correct. Our knowledge of keeping the wild yak is still too slight; it is no longer possible to learn anything of the aurochs, which died out early on. Wild sheep and goats are also gregarious animals for which psychosocial stress in captivity is probably a subordinate factor. Moufflon even seem to be more highly resistant to parasitic diseases when kept in zoos than are other geographically separate forms of wild sheep so that they can be kept without difficulty in conditions that give rise to problems with other wild sheep. This resistance, however, may be connected partly with the fact that the moufflon in question are of Sardinian–Corsican origin and obviously have already passed through an initial stage of selection in the direction of domestication.

In contrast to the related hare, wild rabbits are extremely gregarious, so that caring for and breeding them in captivity is much more successful than keeping wild hares, which is very difficult. Living in groups is also typical of the wild ancestors of the domestic rodents, laboratory rats, mice and guinea pigs. Knowledge of the wild golden hamster is still very limited in this respect. The European common hamster, a possibly comparable species, lives as a solitary animal but can apparently tolerate quite a high population density. However, when young animals of this species are reared in consistently large groups in captivity their growth is very much reduced and sexual maturity is retarded or fails to occur, i.e. the signs of high stress become apparent (Fig. 5.1).

The concept of an increase in psychogenic tolerance due to the decline in environmental appreciation in the course of domestication enables inferences on the general physical development during the transition from wild to domestic animal to be drawn, which can then be compared with archaeozoological finds (Fig. 5.3). Thus hypotheses on the recognition of domestication processes in the material found can be checked for probability. Two extremes of breeding wild animals in captivity can be contrasted with each other: (1) breeding under minimal psychosocial stress, i.e. without any sort of overcrowding or with precisely adequate stocking of an enclosure; and (2) breeding in heavily overstocked enclosures. In the first case, no alterations in the growth of subsequent generations due to psychosocial stress are to be expected, so that there would be no reason for selection factors to intervene in the tolerance system. This method corresponds to the existing situation in well-managed zoos, where the animals are kept in conditions suitable to their

Fig. 5.3. Theoretical scheme of the different paths of development of the growth and the stock of animals bred in captivity over many generations to be expected under various conditions of husbandry and selection procedures (only the extreme situation is depicted in each case). The difference between the process of domestication and mere breeding for conservation in zoos shows clearly.

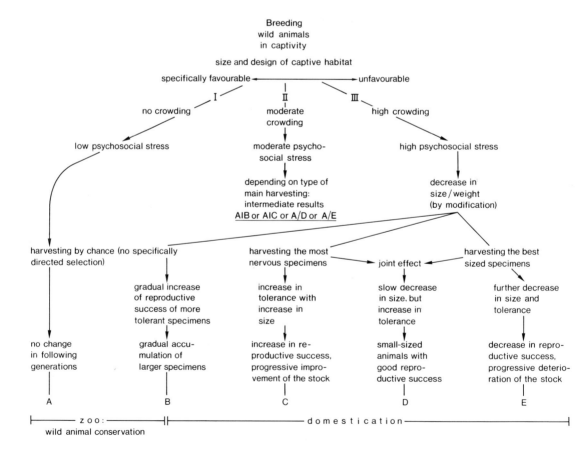

species. The second extreme method is only possible with species that possess the corresponding basic requisite for domestication, in other words which still manage to reproduce, even if only to a limited extent, when kept closely confined in groups. The course of breeding in captivity in all other species can be classified between these two extremes, the capacity for being bred increasing with increasing psychogenic tolerance.

Besides leading to a reduction in the reproductive performance and so to a fundamental slowing down of the overall breeding process, breeding in overstocked conditions repeatedly leads to a reduction of body size and body weight (weight, of course, being also a measure of size) during the growth phase of subsequent generations. A captive population built up in this way, which will possess a certain variability in environmental appreci-ation and psychogenic tolerance and consequently in the physical development of the individual animals, can form the point of departure for three different main directions of selection not initially aimed at breeding to produce alterations.

First, the largest, heaviest specimens may be removed either because they provide the largest amount of meat for consumption or because they are the most imposing for purposes of being passed on as gifts for example. As a result of this selection, there would be a further decrease in the mean size in addition to the reduction in size caused by modification, i.e. by non-hereditary factors, as a result of stress. Insofar as the differences in size were based on differences in psychogenic tolerance, the mean tolerance level of the population would be progressively lowered in this way. This must lead ever closer to the point where the psychosocial stress on the individual animals becomes so great that successful reproduction can no longer take place. In this way, therefore, the reproduction rate would be continuously lowered and the stock would deteriorate more and more.

Second, the most restless, the most timid or the most aggressive animals may be removed either because they are the most likely to escape captivity of their own accord or because herdsmen or keepers choose them first when it is a matter of taking animals out of the stock or slaughtering them, since they cause the most trouble. Insofar as these behavioural differences are in turn based on differences in environmental appreciation (on psychogenic tolerance), such selective measures would be accompanied in the long run by a medium increase in tolerance, a decline in environmental appreciation, and the size reductions due to stress would also be decreased. This means body size will tend to return to its original value when the animals were first taken into captivity for breeding, and indeed may even go beyond it. At the same time the reproduction rate will be raised and so the stock slowly improved on the whole.

In the prehistory and early history of domestic animal keeping, a mixture of these two partly contradictory forms of selection, i.e. the escape or early slaughter of the particularly lively animals and also the removal of particularly imposing ones, has to be considered. On the one hand, such a process would probably lead in the end to a reduction in mean body size, as it involves the concurrent selection of body-size diversity factors not connected with stress. This would however, be distinctly more slow than selection purely according to the first type of selection. On the other hand, it would probably lead to an accumulation of more highly stress-resistant individuals with a reduced environmental appreciation, though also more slowly than according to the second, behavioural type of selection. The final result of such mixed selection would be a stock of small animals with good reproductive performance.

A third possible main direction of further stock management can be described as the lack of any purposeful selection. It occurs when animals are removed from the herd purely by chance. The most that can be expected to happen as a result is a gradual general increase in diversity instead of a shift in a particular direction. The somewhat higher reproduction rate of the psychosocially more highly tolerant individuals will cause the proportion of large animals in the whole stock to rise. This is the other extreme arising from man's non-selective breeding of particular species in zoos. In the practical context of early domestication, it is probably insignificant.

For the time being, it is only possibly to gain experimental and direct observational confirmation for such qualitative changes in individuals in a population of natural high density or bred in captivity. Rodents, which have a certain fundamental level of psychogenic tolerance, rapid growth and rapid sexual maturation, and in which numerous generations follow each other rapidly, are very suitable for this purpose. Using a North American species of vole with a three to four year cycle of population density, Barry L. Keller discovered that, at each peak of the cycle, the population is characterized by heavier and larger animals that grow very rapidly. European field voles were bred in captivity at the Agriculture College at Brno, Czechoslovakia, following the theoretical line discussed here. Initially, the animals kept in cages were of a very small mean size which, however, increased from generation to generation until finally quite large animals were produced.

Archaeozoological studies of the changes in the size of domestic cattle in Britain and in Central and Eastern Europe from the beginning of artificial breeding suggest the picture expected for the case of mixed selection. From the Neolithic, when the shoulder height of cattle fluctuated around 125 to 130 cm, the body size became smaller towards Roman times. By the Bronze Age in some regions there were already dwarf cattle only about 110 cm high. This developmental trend was then interrupted by the great ability of the Romans in breeding livestock, which took effect in a distinct increase in size. The loss of the knowledge of proper breeding measures towards the Middle Ages led to some renewed decrease in mean size that was finally reversed again by planned breeding at the start of the modern age, when improved cattle breeds far exceeded the Neolithic and Roman cattle (Fig. 5.4). The picture outlined here, however, reflects only the mean situation (taking into consideration the various local breeds) and only for specific geographical areas. In fact, there has been an increasing diversity of body size in regional breeds at any one time, the longer cattle keeping has continued. In Germany alone, there are today shoulder heights of between 115 and 145 cm and weights of between 350 and 900 kg amongst the cows of various breeds.

A reduction in size during domestication makes it possible to distinguish between archaeozoological finds of remains of wild and domestic animals from their size alone with increasing certainty. However, it would be a mistake to want to draw in every case the conclusion that the smaller the size difference between domestic and wild animal bones at an archaeological site the shorter is the time since domestication occurred. As the example of cattle shows, breeding to achieve increased size can indeed lead to a renewed size compensation many millennia after the time of domestication.

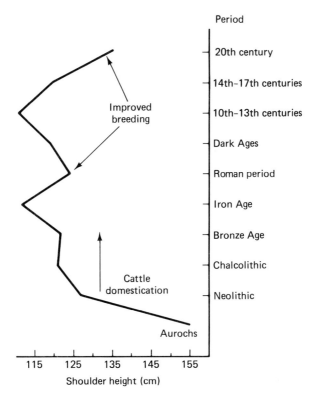

Fig. 5.4. Average changes in the size of domestic cattle in the ages since their domestication from the aurochs (Poland and Hungary). (Re-drawn from Bökönyi, S. (1974), *History of Domestic Mammals in Central and Eastern Europe*. Budapest: Akadémiai Kiadó).

Synopsis

Under the same external conditions, the genetic decline in environmental appreciation that determines the behaviour of domestic animals leads to increased psychogenic tolerance as compared with wild animals, i.e. to less stress. Stress is the state of general activation of an organism brought about by the influence of stimuli; it is a function of the stimuli impinging on the individual and the information gained from them, which is slight in the domestic animal. The seeming contradiction between a general attenuation of behaviour in domestic animals and the sole intensification of sexual activity is resolved by the stress concept. Since psychosocial stress resulting from living in a group plays a central role, wild animals that can be kept in groups in confined conditions are the most suited to domestication. Over the course of numerous generations, influences on growth and reproductive success, tending in different directions and arising from different keeping conditions and different selective measures applied by the keeper, lead to different stock developments typical of wild animals kept in zoos or of domestication.

6 | Acquisition and processing of information

As discussed in the previous chapter, stress depends on environmental appreciation; that is, on which and how many stimuli impinge on the individual and on what and how much information is finally extracted from processing these stimuli. Consequently, the stress level depends in the first instance on the quality of the stimulus-receiving system, i.e. the sensory organs. Furthermore, it depends decisively on the quality of the information-processing system, i.e. the memory capacity and complexity of circuitry of the brain.

Differences in both the sensory organs and the brain are found when wild and domestic animals are compared. As a general rule these changes indicate a qualitative deterioration in the systems for information acquisition and processing and so show the basis of the decline in environmental appreciation. Comparative studies on sensory organs deal mainly with the eye and ear. Studies on the retina and the optic nerve of wolves and of poodles, using light and electron microscopy, carried out by Lothar Schleifenbaum, revealed a decrease in the neuronal portions of the poodle's visual organ, i.e. in the sensory cells, ganglion cells and fibres in the optic nerve. There are even differences within individual breeds. Schleifenbaum found that the light-reflecting layer at the back of the retina, the tapetum, which in many mammals causes the apparent nocturnal 'glowing' of the eyes in a beam of light and increases their visual acuity in weak light, is missing in an inbred line of large poodles though it is present in miniature poodles.

Compared with wild pigs, domestic ones have a much smaller number of rods in the retina, so there is a reduction in the system concerned with light–dark vision. In addition, domestic pigs seem to be short-sighted. Surprisingly, the weight of the eyes of laboratory rats is higher than those of brown rats, but what structural alterations this involves have still to be ascertained. Finally, the eye of the ferret is smaller than that of the European polecat.

Many striking changes to the ears of domestic animals can be observed even externally. Lop-ears and ears with pendulous tips are familiar in the dog (Fig. 2.18, p. 25), but lop-ears are also characteristic of single breeds or groups of breeds of other domestic animals such as pigs and goats. This destroys the ears' normal function as sound receptors; the external auditory canal is deformed, which must result in a reduction in the performance of the sound-collecting system. The dog has a smaller tympanic membrane and smaller auditory bullae than the wolf; its sensitivity to high frequencies seems to be lower and its auditory threshold higher. In wild cats of the genus *Felis*, the size of these auditory bullae is related to the habitat of the species or subspecies. It increases greatly from humid to dry, from thickly vegetated to sparsely or scarcely vegetated habitats, and from swampy regions to sand dune regions. Domestic cats do not fit into this sequence, their auditory bullae being distinctly smaller than those of the African wild cat. The bullae of the ferret are reduced compared to those of the polecat. A reduced auditory efficiency can also be inferred from corresponding differences in the skulls of the guanaco and the llama. Differences also occur in the area of the olfactory organ. A reduction of 11% was found in the area of the olfactory nerves when ranch foxes were compared with wild foxes.

The decisive area for the memory capacity and complexity of interconnections of the mammalian brain is the neocortex, which, in the course of evolution, originally had mainly the function of the rhinencephalon in processing olfactory information. In addition to this function, it then gradually developed into a site for the central evaluation of the stimuli transmitted to it from the sensory organs via the diencephalon. In this way, the cerebrum originated. At the tips of each of its two hemispheres, the olfactory bulbs were separated off as olfactory areas retaining the original function. The basal ganglion responsible for executing innate activities formed at the base. From the reptilian evolutionary stage onwards, the upper wall of the cerebral hemispheres extended to envelop the lower parts of the forebrain as well and so was named the pallium (literally, mantle). This can be divided into the so-called archipallium, an area situated towards the centre of the brain, and the palaeopallium at the

Fig. 6.1. Schematic representation of the evolutionary development and relative positions, up to the mammals of the vertebrate forebrain areas mentioned in the text. The arrows in the cross-sections of one hemisphere for each stage indicate the shifts in position and directions of expansion, the arrows between the pictures indicate the evolutionary steps (based on figures in Romer, A.S. (1959). *Vergleichende Anatomie der Wirbeltiere*. Hamburg, Berlin: Parey). The position of the relevant cross-section within the whole brain is indicated for the primitive mammalian stage.

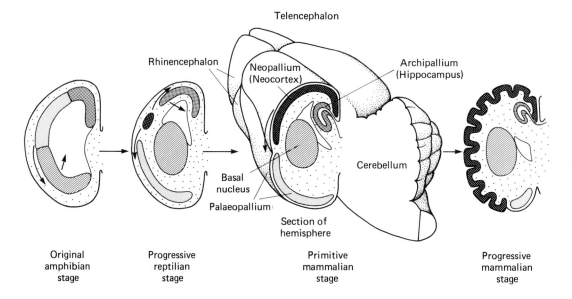

sides. In the reptilian stage, the starting point of mammalian evolution, the neopallium was formed between these two areas. It is also known as the neocortex because it comprises six layers of nerve cells extending over the brain surface. The neuronal pathways from all sensory areas terminate in it and, at the same time, the control of motor activity proceeds from it. In addition to the insula, the visual centre, tactile centre, auditory centre and motor cortex can be considered as primary cortical areas from which coupling pathways lead to all other parts of the brain. Apart from these primary zones, new cortical areas increased in extension during the course of evolution. These areas have mainly integrating functions and so can be called integration or association centres (Fig. 6.1).

The decisive components of this circuitry are the nerve cells or neurones, which are connected to each other via synapses, the junctions between them. In the human brain, each neurone may be connected to up to 10 000 other neurones. Comparing various species of monkeys within the order Primates shows that the number of branches from single neurones, i.e. the complexity of interconnections, increases as the size of the brain increases. At the same time the absolute number of neurones increases but the number relative to the brain size decreases; that is, the neuronal *density* drops.

To find room for complex integration systems such as the neocortex, with an extensive surface area, when the surface becomes too large in comparison to the space available requires a greater or lesser degree of folding, depending on the ratio of the surface area to the space available (Fig. 6.2). When comparing species, it is found that large brains have more folds and grooves than small ones. The furrowing of the neocortex in mammals is related to the total brain volume. As the brain size increases, the surface of the neocortex becomes almost proportionately

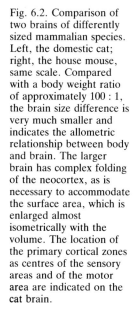

Fig. 6.2. Comparison of two brains of differently sized mammalian species. Left, the domestic cat; right, the house mouse, same scale. Compared with a body weight ratio of approximately 100 : 1, the brain size difference is very much smaller and indicates the allometric relationship between body and brain. The larger brain has complex folding of the neocortex, as is necessary to accommodate the surface area, which is enlarged almost isometrically with the volume. The location of the primary cortical zones as centres of the sensory areas and of the motor area are indicated on the cat brain.

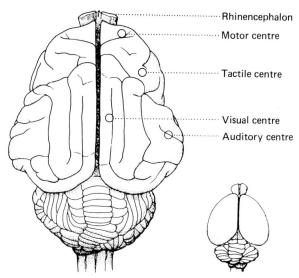

Rhinencephalon

Motor centre

Tactile centre

Visual centre

Auditory centre

larger; the total length of the system of furrows increases both absolutely and relatively. The volume of the neocortex is also similarly empirically related to the size of the whole brain. So, at least for groups of species related to each other, a basic statistical estimate of brain complexity can be obtained from the size or volume of the whole brain, an easily measured value. This estimate is then a rough measure of the fundamental efficiency of a mammalian brain (Fig. 6.3).

Comparing absolute brain sizes, however, makes no sense when comparing animals of different body size, since there is a fundamental interdependency between brain size and body size that can be expressed by the allometry formula,

$$\text{brain size} = \text{integration constant} \times \text{body size}^{\text{ allometric exponent}}$$

(generalized as: $y = bx^a$). As the allometric exponent is to a large extent in the same range for all mammals, the specific integration constant for a species or population, which may be designated the brain size value (cephalization constant, see e.g. Fig. 3.4 (p. 39)), can be determined from the brain and body size of adult individuals of a population. The brain size can be expressed as brain weight or as brain volume, the body size as body weight or a measure of body or skull length. In comparing wild and domestic animals, a certain caution is pertinent for the reference pair brain weight/body weight, as the stress concept indicates the possibility of higher body weights in domestic animals than in the otherwise equally large respective wild forms. This makes spurious differences a foregone conclusion and leads to a domestic animal seeming to have a somewhat smaller brain relative to its body weight, even when the true brain size is the same. Various measurements of length are available as points of reference for comparison of the braincase volumes. Increased

Fig. 6.3. Surface area and degree of folding of the cortex have an empirical relationship to the brain volume (as a measure of brain size) in mammals, which makes it possible roughly to infer the complexity of a particular brain and to use the size of the brain as an approximate measure of its fundamental efficiency.

(*a*) Cortex surface plotted against brain volume. (*b*) Ratio between cortical surface area and length of exposed gyri. ○, marsupial; ●, carnivore; △, human; ▲, whale. (Redrawn from Elias, H. & Schwarz, D. (1971), Cerebrocortical surface area, volumes, lengths of gyri and their interdependence in mammals, including man. *Zeitschrift für Säugertierkunde*, **36**, 147–63.)

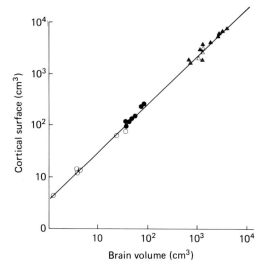

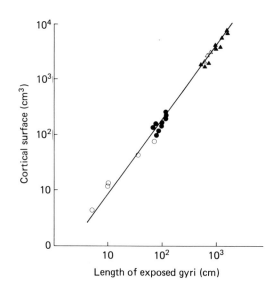

caution has to be maintained for measurements of the overall length of the skull when there is a possibility of different muzzle lengths entering into the overall measure, as shortened muzzles are frequently found in domestic animals. Only measuring the base of the cranium (e.g. basicranial axis) provides more certainty though this value cannot be treated as absolute because an alteration in the proportions of this area cannot be excluded either. Consequently, both weight and length should be taken into account to supplement each other whenever possible, especially if there are only slight differences in the brain sizes in a comparison between wild and domestic animals or if a more exact determination of percentage differences is intended. Exact statements of percentages must remain somewhat questionable in any case.

There are differences between the relative brain sizes of geographically separate populations of wolves (see Chapter 3: Figs. 3.4 (p. 39) and 3.7 (p. 43)). The wolves of the northern regions of Eurasia have the highest values, North American wolves possibly somewhat lower ones, and the wolves of the Arabian Peninsula and south Asia (the southern wolves) the lowest. The brains of the last group are at least between 5% and 10% smaller than those of the northern wolves (Fig. 3.4). The brain sizes of domestic dogs are even smaller. Dingoes, extremely primitive dogs that returned to the wild early on, have brains about 25% smaller than those of northern wolves; brains of primitive dogs from the tropics and East Asian breeds are another 15% to 20% smaller than that of the dingo. Most European improved breeds take a position intermediate between these lowest levels and the dingo level, although their mean value comes very close to that of the dingo. The small pinscher breeds rank very low, the terriers somewhat higher. Greyhounds and some watchdog breeds sometimes exceed even the dingoes and approach the level of the southern wolves. However, there are wide ranges of scatter in the brain sizes of both single wolf populations and single dog breeds, and two-thirds of the animals in any one of them fall within the range of ± 10% around the mean for each. So there are broad areas of overlap between the ranges of variation of, on the one hand, southern wolves and dingoes or large-brained improved breeds, and, on the other, dingoes and the primitive dogs with the smallest brains. Since the dog appears to derive primarily from the southern wolf, there must have been a reduction of its brain size during its domestication through the dingo-type stage to the rest of the primitive dogs. It cannot be ruled out that the dingo's brain size has even increased, compared to that of the most primitive dog grade, during the millennia of its living in the wild. Measuring the braincase volumes of European primitive dogs shows a corresponding smallness of brain from the Neolithic up to the Roman Age and then to the Middle Ages. A renewed increase in brain size must then have taken place during the development of the modern European improved breeds, which has led to the present mosaic of differences between breeds. Such differences have been demonstrated even as early as Roman times, when dog remains from German coastal sites had larger cranial volumes than those in the Rhine–Main area.

There are also differences in brain size between the different geographically separate populations of wild cats. The brain is largest in European wild cats and smallest in the wild cats of northeast Africa and Palestine, and in the wild cats of Southwest and South Asia. Other African wild cat populations and the wild cats of Soviet Central Asia are intermediate between these two extremes and come close to the size of the European populations. As in the wolf, it turns out that in this case also the domestic form descends from the population or population group of the wild species with the smallest brain. Ancient Egyptian cats, of which a large number of mummies have been preserved, still had the same brain size as their wild progenitors. As far as it known, a reduction of about 10% is not detectable until the Middle Ages, in cats corresponding to the modern European domestic cat. Finally, a further reduction in brain size of about 5% to 10% is to be found in Siamese cats.

In general, compared with the European polecat, the ferret has a much reduced brain (21% for the braincase volume/condylobasal length ratio and 29% for the brain weight/body weight ratio). This characteristic has not yet, however, been studied in the polecats of the Mediterranean region that are evidently the initial group in the domestication process.

Both primitive and improved breeds of horse have smaller brains than the only wild horse still extant, the Przewalski horse, though thoroughbreds just attain the latter's brain size. Nevertheless, since the Przewalski horse is not the source for primary domestication, any inference drawn from its measurements is not reliable. There are scarcely any measurements of the brain volume of truly wild horses from Europe and none from those of eastern Europe. For these, we have to refer to finds from the Upper Pleistocene and first Postglacial millennia as it is not possible to distinguish clearly between skulls of wild and domestic horses from any later archaeozoological material. The skull of one of the last Russian tarpans has the same brain size as that of a domestic horse but an admixture of domestic horse in the tarpan population cannot be ruled out, so that there can be no certainty here either. Possibly there was really not much reduction of brain size in horse domestication, which may have started with a population that may have had smaller brains than the Mongolian Przewalski horse. A late Pleistocene English wild horse cranium also yielded a volume in the middle of the range of those of domestic horses, as did some skulls of uncertain but probably early Postglacial age from the Rhine–Main area of Germany.

The brain size of the donkey is clearly smaller than that of the wild ass.

Considerably different brain sizes are found in the species and geographically separate populations of wild pigs that formed the basis of the domestic breeds. The wild pigs of Europe and the Near East have the largest brains, the next largest probably being those of populations from northern and eastern Asia. Southeast Asian banded pigs have 20% smaller brains that the latter, while Indian ones are intermediate. Warty pigs are at about the level of banded pigs (Fig. 3.24, p. 57). So pig domestication evidently also began with the forms that had particularly small brains, as has already been ascertained for the dog and the cat. The Papua pigs, probably the most primitive of all domestic breeds from the transitional zone between wild and domestic animals, do not differ from the warty pigs in this respect. Primitive domestic pigs from the Sudan have brains about 25% to 30% smaller than those of warty pigs and banded pigs. The Chinese masked pig, an improved breed, has a similarly small brain. By contrast, though the brains of European landrace pigs are around 30% smaller than those of European wild pigs, they are only a little smaller than those of warty pigs and banded pigs. Here an influence of the massive secondary domestication of European wild pigs, i.e. their crossbreeding into early primitive breeds of domestic pigs, has to be taken into account. In a comparative analysis of single brain areas of European wild and domestic pigs, Dieter Kruska and Heinz Stephan demonstrated a heavy reduction in the centres connected with the sensory organs, the visual centre being the most affected.

For camels, it is impossible to compare the dromedary with its unknown wild counterpart. Its brain, however, is still smaller than that of the Bactrian camel, which in turn has a distinctly smaller brain than the wild camel from the Gobi Desert. There is a broad overlap between the llama and its ancestral species, the guanaco, but the average brain size of the llama is about 15% smaller. The brain of the alpaca is about 5% smaller than that of the llama and that of the vicuna, these being approximately the same size.

Brain sizes have not yet been compared in domestic reindeer and their wild ancestors, the Eurasian reindeer. Feral Chillingham cattle show a brain size reduction of between 10% and 20% from that of Pleistocene and Postglacial British aurochs (Fig. 6.4). Archaeozoological material from the Rhine–Main area indicates that the brain sizes of early Central European domestic cattle had the same range of scatter. A certain reduction in brain size also took place during the domestication of the yak. There are as yet no comparative studies for other kinds of domestic cattle.

Different geographically separate populations or population groups of wild sheep typically have differently sized brains. The European mouf-flon, obviously having originally passed through an initial stage of farming, has a particularly small brain, while that of the true originally wild moufflon from the Near East and the urial group is somewhat larger. The highest values are found in the argali from the high mountains of Central Asia. So domestication began with a small-brained group in this case also. Though there are distinct differences between breeds, domestic sheep have brains smaller than that of the moufflon. While the reduction in the Hungarian Zackel sheep is only slight, it is of the order of around 20% for the extremely primitive Soay sheep. Among the local breeds, the German heath sheep also ranks at this level.

The brain size of the domestic goat is smaller than that of its wild ancestor. Comparative studies are still in their initial stages, but first indications are that eastern bezoar goats, including the Pakistan Chiltan Hill population in the *aegagrus* and *falconeri* type border zone, have small brains.

Brain size studies carried out by Heinz Moeller on domestic rabbits illustrate the problems involved in using only the brain weight/body weight pair for domestic animals. Different directions of selection created large differences in growth types with very different weight levels in many breeds of domestic rabbit. Breeding to produce meat led to high weights in such 'commercial breeds' as the European lop and the Viennese White in contrast to the low weights of the 'sport breeds' such as the Belgian lop and English lop. When body weight is the only basis for comparison, differences in brain size must be treated with caution. The average for all breeds would result in an apparent reduction by 24% from the wild rabbit, but the reduction amounts to only 16% for the domestic rabbit with the body build closest to that of the wild form.

Brain size reductions from that of the wild ancestor range between 5% and 10% for domestic guinea pigs and for laboratory rats. No significant general reduction in brain size has so far been discovered between laboratory and wild house mice.

As this review suggests, domestication of mammals as a general rule involved a fairly considerable reduction in the brain size. That the neocortex would have to be affected most due to its general relation to the whole brain has been directly confirmed in all species on which quantitative studies of individual brain areas have been carried out. The

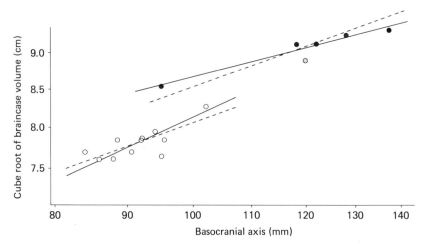

Fig. 6.4. Scatter diagram comparing skull braincase volumes of English Pleistocene and Postglacial aurochs (●) with that of today's feral Chillingham cattle (○). Unbroken lines, allometric lines as actually calculated; broken lines, parallels indicating the probable mean transposition between aurochs and Chillingham cattle; circle, another aurochs skull that could not be measured with full accuracy due to its broken state.

reduction in storage capacity and complexity of interconnections must without doubt affect information processing in this system. Changes in information processing in turn correspond to a change in environmental appreciation that is fed with all kinds of stimulus (cf. Chapter 5). Indeed, a close correlation was found between brain size and the stress indicative zona fasciculata of the adrenal cortex in a pilot study with fully comparable young male fallow deer in the domestication program described in Chapter 12. In the special case of reduced performance in the domestic animal, this must mean a reduction of stress as was initially ascertained in behavioural studies. So the structural alterations to the brain and sensory organs during domestication can be interpreted as the morphological results of a selection for decline in environmental appreciation, attenuation of behaviour, and lower reactivity.

How these differences in information processing between wild and domestic animals, which are of significance primarily in different stress levels, are expressed in clearly quantifiable differences in activity, particularly learning performances, is still largely unresolved. Testing is difficult because the domestic animal is less timid; its reactivity is lower and so it is less easily distracted by secondary stimuli. In learning or training, this provides it with an initial advantage over the wild form in recognizing a task or problem, the exploratory behaviour of the latter being inhibited at first.

Intelligence tests with various breeds of dog have so far produced apparently contradictory results with respect to brain size. There have always been large individual differences in performance. In the author's research group, Ruth Vierengel carried out an experiment on dogs, in which a piece of food attached to a cord lying loosely on the ground could be acquired only by pulling on the cord (Fig. 6.5). Boxers, a breed with a comparatively large brain, attained somewhat better results on the average than alsations, and these seemed in turn to be slightly superior to chows, a breed with a relatively smaller brain. Some northern wolves tested by Bernhard Grzimek with the same or a very similar method solved the task at first go. In the case of one female wolf, however, he had to withdraw quite a long way before she began to show any interest at all in the piece of food. This illustrates clearly the problems in comparing domestic and wild animals due to the greater timidity of the latter in such situations. Similarly constructed tests with other breeds of dog, however, led to rather contradictory results. Influences on the relevant performance arising from the different temperaments of the dogs and hence different responses in the test situation (obstructing the direct possession of the piece of food by tying the dog up or placing a grate between it and the food) seem to play a major role. Harry Frank used the same set of tests on American wolves and sledge dogs. The wolves achieved much better results in problem-solving tasks than the dogs but dropped far behind them when, in purely training tasks, the behaviour to be learned

Fig. 6.5. Three phases of an intelligence test on a dog.
(*a*) The dog has noticed the piece of food attached to the end of the cord and strains on its lead without being able to reach the food directly (young Canaan dog, a breed derived from the Israeli primitive dogs).
(*b*) After unsuccessful efforts and pauses of apparent uninterest, the dog has found the solution and tugs at the cord.
(*c*) The goal has almost been achieved.

(*a*)

(*b*)

(*c*)

bore no clear functional relation to the range of stimuli. This apparent lower behavioural flexibility in the wolf may be explicable in the sense that its higher problem-solving intelligence might serve to obstruct its insight into seemingly senseless tasks (but see p. 117 for reduction in behavioural flexibility due to increased dopaminergic activity). A parallel to this is to be found in the fact that trainers of dogs for Customs and Excise prefer less intelligent animals who, after suitable training, stubbornly persist in situations which seem hopeless in relation to their own capabilities.

Certain differences in learning speed and memory performance, which may be connected with body size, have been demonstrated between different breeds of rabbit. This could most likely be interpreted in terms of different rates of metabolism.

Dogs highly bred for special performance, such as watchdogs or sheepdogs, have distinctly larger brains than primitive breeds. The same seems to be the case for European thoroughbred horses. In such cases, where learning ability or memory acquired positive selection value, the original process of brain reduction during domestication has been reversed.

This chapter's review of the brain size situation of the various species of domestic animals compared with their wild ancestors is not simply an illustration of the phenomenon of brain size reduction occurring as a general rule during domestication. It also shows that, in species with distinct differences in brain size between geographically separate populations, domestication began in each case with just that population having the smallest brain size within its species. According to what is known at present, this holds at least for the dog, the cat, the pig, the sheep, and apparently also for the horse and goat. The selection taking place in the course of domestication was, therefore, already at work in the primary acquisition of the animals on which domestication was based. This discovery allows a second principle of domestication to be formulated which, in combination with the first (p. 96), describes a further facet of suitability for domestication:

The individuals with a particularly small brain size in a species to be domesticated seem to be especially suited as a group for the initial domestication process.

Synopsis

The quality of the systems for information acquisition and processing responsible for psychogenic stress via the diversity of environmental appreciation changes in the expected direction towards a decline in environmental appreciation in the domestic animals during domestication. Reductions in the information acquisition system (the sensory organs) affect the eye, the ear and the olfactory organs to varying qualitative and quantitative degrees. In the information-processing system (the central nervous system), the neocortex of the cerebrum is the decisive area for memory capacity and complexity of interconnections. Since its development is connected with that of the whole brain, even measurements of the whole brain provide insight into the relative efficiency of this system, in addition to what can be learned from studies of special cortical areas. Comparing wild and domestic animals usually reveals decreases in brain size that may have been of a considerable extent during domestication. Where there are distinct differences in the brain sizes of a wild species, domestication began right from the start with the populations having the smallest brain sizes. So the smaller the brain size of individuals from a species to be domesticated, the more suitable they seem to be for this purpose.

7 | Transmitter substances for information processing

It is not only the size and structure of the brain, particularly the neocortex, that determines information processing and thus environmental appreciation. The complex of transmitter substances of the nervous system, the neurotransmitters, plays a vital role. Such substances are indispensable for the transmission of excitation at each of the many junctions, the synapses, between the individual nerve cells or the processes coming from them. While the transmission of excitation at the neurone itself is an electrical process, it has to be carried out chemically at the synapses, since it cannot be relayed directly across the gap. As the neural pathway of each nerve is one-way, a presynaptic and a postsynaptic element can be distinguished at each synapse. The presynaptic element is the end of a nerve process relaying the excitation to the synapse; the postsynaptic element is the start of the nerve process relaying the excitation to the next nerve cell. The synaptic cleft is situated between these two (Fig. 7.1).

There are large numbers of synaptic vesicles each containing neurotransmitter molecules in the presynaptic element. These vesicles are activated by the arrival of excitation in the form of a nerve impulse, a change of the electrical potential of the cell membrane (action potential), and move to the cell membrane, where they release the transmitter into

Fig. 7.1. Diagram of the principle of excitation transmission at the synapse. For explanation, see the text.

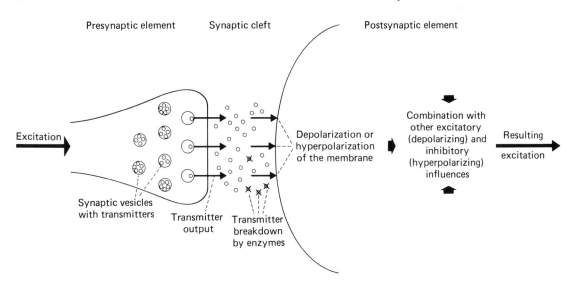

Presynaptic element Synaptic cleft Postsynaptic element

Excitation

Depolarization or hyperpolarization of the membrane

Combination with other excitatory (depolarizing) and inhibitory (hyperpolarizing) influences

Resulting excitation

Synaptic vesicles with transmitters

Transmitter output

Transmitter breakdown by enzymes

115

the synaptic cleft. The postsynaptic membrane possesses specific receptors each structured to receive a certain neurotransmitter. These receptors mediate the effect of the transmitter on the postsynaptic membrane so that local discharges occur. Such 'excitatory' postsynaptic potentials can be added together. When the potential increases beyond a certain threshold, it triggers an action potential which corresponds to the relay of the excitation. The summation of the postsynaptic potentials can be either temporal or spatial. Temporal summation refers to the impulse frequency at the presynaptic element, which provides information on the excitation amplitude to be relayed in the form of the 'impulse frequency' code of the transformation of a stimulus into excitation. Spatial summation is based on the computation of the influences of several neurotransmitter streams at several synapses having affect simultaneously upon the same neurone.

Not all neurotransmitters cause discharges, i.e. depolarizations, of the postsynaptic membrane, some can also, depending on their specificity, produce hyperpolarizations, i.e. further charges; so the postsynaptic element is really the computational unit of the neurone. A new action potential and consequently a relay of excitation does not take place until the sum of the depolarization exceeds the sum of the hyperpolarization by a certain threshold. In this way, the specificity of the neurotransmitters from excitatory and inhibitory neurones, and their quantitative interplay at the postsynaptic elements of the synapses, control the overall information processing in the nervous system. Enzymes in the synaptic cleft and at the two synaptic elements provide a rapid deactivation of the transmitters released into the synaptic cleft. This multifactorally controlled, and so rather labile, system offers an easy starting point for various ways of influencing the transmission of excitation, and thus the whole of information processing, by means of numerous psychoactive drugs which can affect the synthesis and storage of transmitters in the presynaptic element, the release of the transmitters and their enzymic breakdown, or the receptors of the postsynaptic membrane (Fig. 7.1).

Among the many substances with neurotransmitter properties, acetylcholine and the so-called biogenic amines seem to be especially important. The indolamines (5-hydroxytryptamine) and catecholamines belong to the latter, a group which seems to be of special interest for the understanding of the domestic animal phenomenon. They are synthesized in a joint biochemical path starting from the amino acids phenylalanine and tyrosine via the intermediate product dopa in a progressive sequence to dopamine, noradrenaline and adrenaline (Fig. 7.2). Dopamine and noradrenaline play a central role as neurotransmitters, adrenaline being particularly important as the hormone that generally activates the body and is secreted from the adrenal medulla upon acute stress, i.e. as a direct consequence of the influence of 'stressors' (fight and flight syndrome, see p. 94). Its function as a neurotransmitter seems to be secondary.

Increased dopamine secretion in the brain is accompanied by strengthened reactions, intensification of activity, particularly in the aggression complex. Increased dopaminergic activity seems to involve a reduction in behavioural flexibility to the point of stereotyped repetition of action sequences. Such behavioural changes may be traced to an increase in sensory perceptions or in the significance of stimuli otherwise a matter of indifference for the organism, i.e. back to a corresponding change in information processing. Dopamine appears to have a central significance in the occurrence of all major, endogenous psychoses such as manic-depressive diseases and schizophrenia that involve functional disturbances in the area of information processing. Psychosocial stress influences the metabolism of catecholamines via shifts in the enzymic activities mediating their synthesis and so affects information processing. Psychopharmacological interventions in the dynamic balance of the neurotransmitter system throw light on the connection with information processing and the occurrence of stress. Amphetamine, for example, causes the secretion of catecholamines and interacts with their storage, uptake, release and deactivation. The rest of its physiological effects are connected with this. According to the level of stress amphetamine leads to an increase in the adrenocorticotrophic hormone (ACTH) of the pituitary gland and to a rise in the corticosteroid level in the blood (cf.

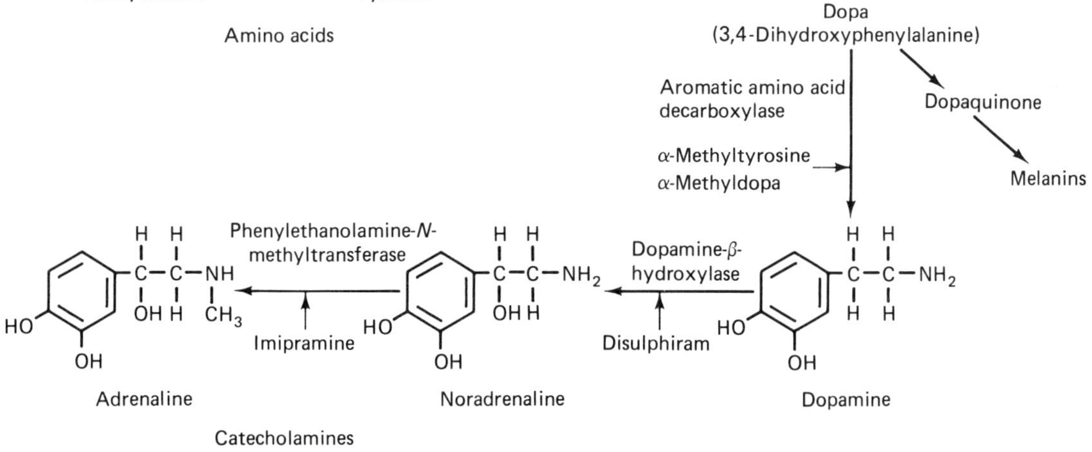

Fig. 7.2. Diagram of the synthesis of the catecholamines from the amino acids phenylalanine and tyrosine. The sites of the branches to the synthesis of the melanins, the coat-colour pigments, and the synthesis of thyroxine, the thyroid gland hormone, are indicated. The enzymes that initiate the relevant synthesis are indicated above or beside the arrows. Inhibiting substances for these reactions are given in italics below them. (Compiled and supplemented from diagrams in Schade, J.P. (1969), *Die Funktion des Nervensystems*. Stuttgart: Fischer.)

Chapter 5). The alterations to information processing caused by amphetamine can be experienced by humans directly: light, sounds, tastes, odours, in brief sensory perceptions seem to be more intense. Similar experiences, though with further-reaching disturbances of the information-processing system, are elicited by hashish (marijuana). ACTH secretion is stimulated and, depending on the dosage, there are changes in mood and strengthening of sensory perceptions ranging up to spatio-temporal estimation errors, illusions and hallucinations – to confusion in environmental appreciation.

These ways of influencing the information-processing system provide another level of experimental verification of alterations in the dynamic balance of the neurotransmitter system, which are postulated as being responsible for the typical changes in behaviour occurring during the transformation from wild animal to the domestic form. The application of tranquillizers used not only in human medicine but also in animal breeding (partly to reduce stress) seems to be especially appropriate for simulating the transformation to the domestic animal. In the author's research group, Heiko Ernst undertook such a model experiment with chronic dosages, i.e. ones administered over several months, of chlordiazepoxide (Librium) in a special feed mixture. This drug affects the catecholamine system amongst other things. The animal selected for the experiment was the North American cotton rat (*Sigmodon hispidus*), a species distantly related to the hamsters and used as a laboratory animal for certain parasitological studies and for evaluating central-depressive substances (Figs. 7.3 and 7.4). It is very easily alarmed and very susceptible to disturbance, which makes it well suited to these purposes. Cotton rats like to have close contact with conspecifics belonging to the

Fig. 7.3. Cotton rat (*Sigmodon hispidus*), the animal used as the subject in a model experiment to simulate the transformation from the wild to the domestic state by chronic dosage of tranquillizer.

same family unit but are extremely incompatible with strangers. Only in situations of social disorganization can animals of different kinship become accustomed to being together in the same group but even then violent fights take place constantly. Even family groups do not remain undisturbed in the long run. From time to time, single animals will be socially isolated and, in a cage, which offers no way out, finally killed.

A standardized jet of air was used as a stimulus for registering changes in the flight reaction after the beginning of the drug treatment; this air jet was directed at the animal's neck from a 2 mm wide opening in a pipette with a rubber ball at the end. Untreated cotton rats reacted to this stimulus first by jumping up into the air in alarm an average of 20 cm and then fleeing for about 3 s for a distance of about 75 cm. After a week of drug intake, the jump no longer occurred at all, the flight duration dropped to about 1 s and the flight distance was only about 10 cm. At the same time, the frequency and intensity of bodily contacts with conspecifics dropped to nearly half of the values in a control group. Half-grown animals tended more to social isolation in the cage corners; their huddling was reduced. The frequency of serious fighting dropped considerably and no more biting was observed after the middle of the second week of the treatment. It was even possible to introduce animals from other groups into an existing group without any problems. Nest defence by mothers disappeared. Normally a cotton rat with young violently attacks conspecifics coming closer than 10 to 15 cm. Now, in extreme cases, the mothers even huddled with other animals on the immediate area of the nest so that young were either crushed to death or pushed out of the nest. The nesting site, usually under as much cover as possible, was now, under chronic influence of chlordiazepoxide, chosen more at random, being situated

Fig. 7.4. Cotton rat mothers kept in captive groups normally defend an area of from 10 to 15 cm around their nest extremely violently against conspecifics. This defensive behaviour was diminished in the domestication simulation experiment in a way similar to the difference between wild and domestic animals.

even in the middle of the cage. The otherwise usual spatial separation of feeding, resting and defecating places gradually disappeared. Young born and growing up in the drug period were about 40% heavier than control litters at the age of five months. The control animals that were kept in groups had litters of four to five young on the average, while the litter size of the test animals rose to six or seven, reaching the litter sizes of mothers kept as part of a pair and so not subject to high psychosocial stress.

Thus, the chronic dosage of tranquillizer brought about a considerable reduction in the readiness to take flight and the alarm reaction, a loosening of social bonds and simultaneously a strong drop in aggressive reactions and so higher social compatibility. This was an attenuation going as far as complete disappearance of differentiated modes of behaviour, particularly care of the young. However, it had no negative influence on reproduction and even increased the litter size and caused the young to grow to a heavier weight. This corresponds completely to the behavioural changes from the wild to the domestic animal summarized in Chapter 4 and is in full agreement with the environmental appreciation – stress concept of the change from wild to domestic animal formulated in Chapter 5. The experiment is therefore a confirmation of the idea that domestication involves a drop in information processing to a new, reduced level of environmental appreciation.

Synopsis

The dynamic balance of the excitation transmitter substances in the nervous system, the neurotransmitters, is of decisive significance for what happens in information processing and so for environmental appreciation. Psychoactive drugs that influence the operational mechanisms of the neurotransmitters at the synapses alter this balance in various ways. This opens up the possibility of experimentally simulating, psychopharmacologically, the attenuation in information processing which has been postulated as the cause of observed alterations in behaviour occurring during the transition from the wild to the domestic state. An experiment of this type carried out with the cotton rat as a model of the wild animal confirmed the concept of the alteration of a complex consisting of information processing, stress and behaviour.

8 | Coat colour and behaviour

The melanins (coat-colouring pigments, cf. Chapter 2) are produced by the same basic biochemical pathways as the catecholamines essential for information processing. The amino acid dopa (3,4-dihydroxy-phenylalanine) synthesized from tyrosine by the action of the enzyme tyrosinehydroxylase is the basis for the further formation of both groups of compounds. Dopamine, the first in the series of catecholamines, is produced from dopa; while the melanins are built up via the intermediate substance dopaquinone (Fig. 7.2, p. 117). There are self-regulating feedback mechanisms controlling the concentration of the preliminary stage, dopa, and leading to fluctuations in the pathways of catecholamine synthesis; similar mechanisms are known for the action of all neuro-transmitters. It is to be expected, therefore, that there is similar feedback regulation of other utilization pathways for dopa in the living system. The conjecture, first expressed by Clyde E. Keeler, that there is a fundamental relation between the supply and use of dopa in the nervous system and the supply and use of dopa for the synthesis of melanin in the skin seems feasible on this basis.

By means of human hereditary metabolic defects, the connection between these two systems, which seem so completely independent of each other at first, has been demonstrated; it is due to a shared component that forms the basis of the relevant synthesis pathway. Phenylketonuria, sometimes called PKU, involves a complex of symptoms: mental deficiency and extremely light pigmentation of the eyes and hair. It is based on enzymic malfunction in the metabolism of phenylalanine that prevents the synthesis of the amino acid tyrosine from the amino acid phenylalanine, which is the basis for the subsequent formation of dopa.

Connections between hair colour and behaviour used to be current in popular lore, as for example among experts on horses: bay horses were characterized as sanguine, chestnuts (horses in various shades of reddish brown, whose manes and tails are not black) as choleric, black horses as melancholic, and greys as phlegmatic. Dog breeders still sometimes regard reddish individuals of some breeds as more nervous and unreliable, and consider the strongly pigmented (dark coloured) ones to be more lively than the light coloured. Finally, popular lore even attributes different behavioural norms to people with hair of different colours. As science has advanced in the twentieth century much of this knowledge, originally based on experience has been lost or rejected and looked upon as unscientific, since the acceptance of such a connection between colour and behaviour seemed completely senseless. But this was before the acquisition of detailed knowledge on the biochemical foundations of both information processing in the brain and pigmentation.

Experimental foundations and quantitative bases for the real existence of such relations between coat colour and behaviour have been carried out since the 1940s, mainly by Clyde E. Keeler in the USA, with studies on laboratory rats, ranch mink and foxes, and in recent years in the author's research group with studies on several other large and small mammalian species. Wolfram Bernhard used psychodiagnostic methods to test 500 soldiers in the German Federal Defence Force and so demonstrated statistically well-founded, psychogenic differences between the extremes of hair and eye pigmentation for man.

Keeler first crossed brown rats with albino laboratory rats that, in addition to the *albino* allele, were homozygotes for the *piebald* allele and for the *non-agouti* (*black*) allele of the *agouti* locus (cf. Chapter 2). The offspring were bred further and the behaviour of the wild-coloured animals (ones in which the *agouti* allele of the *agouti* locus was expressed) and black animals (with the *non-agouti* allele) was compared. The *non-agouti* black rats proved to be less timid and less aggressive and showed more confidence in new situations by exploring them more quickly. Albino rats in these experiments proved to be less ready to react to olfactory stimuli than the others.

Later he observed ranch mink, with a broad palette of coat colours, and measured their bodily development. The dark mink, the standard dark brown animal, was always used as the reference colour. Pastel mink, i.e. the light-brown animals, turned out to be 8% heavier, had smaller adrenal glands and were characteristically less active and aggressive; that is, they behaved more tamely. Silver-blue mink, brownish-grey animals, deviated from the standard to almost the same degree. Black-eyed whites, i.e. carriers of a *white* locus independent of the *albino* locus, were very tame and much less ready to take flight. Palomino mink, pale beige-brownish animals, weighed 14% less and were relatively tame. By contrast, the short, slender Aleutian mink, blackish with a greenish-grey gleam, weighed 5% less, and were typically especially aggressive, as were the grey sapphire mink, which are homozygous for the two colouring alleles responsible for the silver-blue and Aleutian varieties.

Quantitative determinations of behaviour, size development and bio-chemical parameters in ranch foxes were the most significant of Keeler's studies (Fig. 8.1). The animals used were wild red foxes, silver foxes (that is the non-agouti black animals with a *silver* allele); and pearl, a breed colour that combines the *non-agouti* allele with a *dilution* allele ('blue'), amber which carries the *brown* allele of the *black* locus in addition to the two former alleles, and glacier, animals in which a *white* allele is added to those for the amber colouring. Not all measurements, however, were carried out on this last colour so that the full comparison is limited to the wild colour, silver, pearl and amber. The weight of the foxes increased and the relative weight of the adrenal gland decreased in the sequence: wild colour → silver → pearl → amber. There were dramatic changes in the flight distances in the large enclosures where the tests were performed. While silver foxes reacted immediately when approached to within about 180 m and wild red foxes even somewhat earlier, pearl foxes fled when the observer came within about 130 to 150 m and amber foxes put up with an approach to within about 5 m. Their avoidance distance thus amounted to only about a fortieth of that of silver foxes and about a thirtieth of that of pearl foxes. The amount of protein-bound iodine in the blood, a measure of thyroid gland activity, rose in the same sequence by more than 1.5 times and the secretion of catechol acids, catabolic products of adrenaline in the urine, increased clearly. So, in these foxes, the additive combination of the relevant colouring alleles leads to ever more 'tame' behaviour, reaching an extreme in the amber animals, conspicuous by their low activity, low aggressivity and lower reactions to different kinds of stimuli. The connection with the stress system is made clear by the parallel changes in the development of the adrenal gland and its function (adrenaline secretion) and of the thyroid activity. The increase in weight can be understood as a consequence of less stress (cf. Chapters 5 and 7) (Figs. 8.1 and 8.2).

Crossing amber foxes with wild red foxes and back-crossing the

Fig. 8.1. Diagram of the changes in body weight, in the relative weight of the adrenal gland, an essential organ in the stress axis (cf. Chapter 5, Fig. 5.2, p. 94), in the content of protein-bound iodine in the blood as a measure of thyroid function, in the excretion of catechol acids in the urine as a measure of adrenalin breakdown, and in the flight distance from an observer. These variables depend on increasing changes in coat colour through various colour factors (from the results of Clyde E. Keeler).

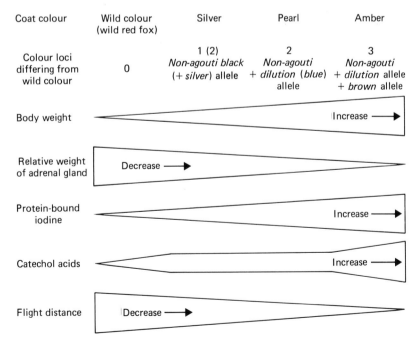

offspring with amber foxes resulted in a range of eight different colour types: red (wild colour), non-agouti black, blue agouti, chocolate agouti, blue non-agouti, chocolate non-agouti, blue chocolate agouti and blue chocolate non-agouti. Of these, the first is not homozygous for any of the altered colour alleles, the second to fourth are homozygous for one each, the fifth to seventh for two, and the last for three of these alleles. In a smaller test enclosure, the average flight distance of the last four of these crossbreeds proved to be distinctly less than that of the first four, being only a little more than half as far. This confirms the feasibility of attaining an increase in behavioural change by adding more of the colour alleles that deviate from the wild colour. As is exemplified by the mink, such

Fig. 8.2. Fox furs of different colours (part of the colour spectrum of Clyde E. Keeler's behavioural and biochemical studies, cf. Fig. 8.1).

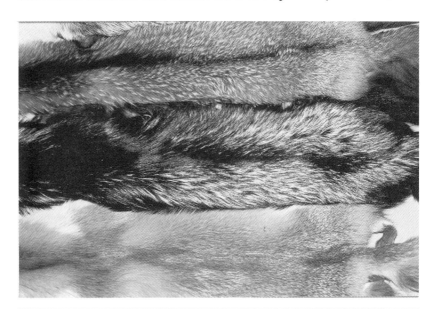

Fig. 8.3. Silver-grey colour type in a stock of black house rats (*Rattus rattus*) (original animal from Horst W. Schwabe's stock).

Fig. 8.4. Shifts in the frequencies of behavioural intensities in a house rat (*Rattus rattus*) group kept for breeding, where the animals differed in only one allele of a colour factor: black and silver-grey animals. These results of Horst W. Schwabe make it clear that behavioural differences involved in changes in coat colour can be indicated statistically very well when there are large numbers of observations and experimental animals, but, due to wide scatter, this does not necessarily make it possible to classify individual animals. For further explanation, see

(*contd on p. 126*)

an alteration need not always mean an increase in 'tamer' behaviour.

Horst W. Schwabe tested and compared silver-grey animals that occurred in his stock of black house rats (*Rattus rattus*) with the black ones (Fig. 8.3). In general, he observed more nervous, sensitive reactions in the silver-grey rats, tending to more violent flight reactions and to stereotyped behaviour, i.e. multiple repetitions of identical action sequences indicating restlessness. Their overall locomotor activity (motility) was higher in a large arena but the number of times they stood upright, a part of exploratory behaviour, was lower. They explored an unknown room less intensively than the black rats. When directly harassed by the observer they tended to attack more quickly, but were less aggressive towards conspecifics, towards which they behaved more indifferently; after having been accustomed to being together with other unknown animals, they huddled less with these than the black animals did. So, on the whole, the colour changes from black to silver-grey was connected with higher motility, generally more restless behaviour, lower exploratory activity and higher social indifference (Fig. 8.4). Finally, the

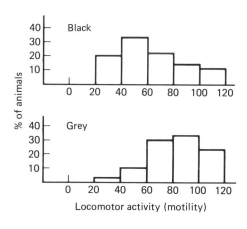

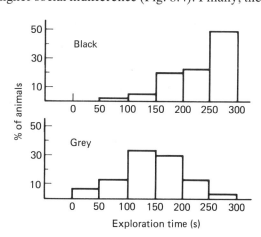

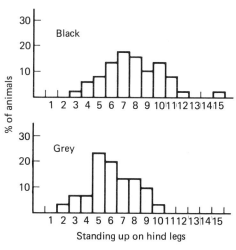

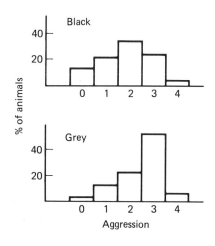

the text. (Redrawn from figures in Schwabe, H.W. (1979), Pigmentationskorrelierte Verhaltensunterschiede bei Hausratten (*Rattus rattus* L.). *Zoologisches Jahrbuch, Systematik*, **106**, 406–26.)

silver-grey animals were about 10% lighter in weight than black ones of the same age. These behavioural changes are reminiscent of those observed in rodents under the influence of amphetamine. Though small doses of amphetamine increase motility, standing upright on the hind-legs, exploration and social interaction, higher doses reduce all of these and, at the same time, more stereotyped behaviour occurs. These parallels as with reactions to amphetamine seem to indicate a higher secretion in the catecholamine system of the brain of silver-grey rats.

Differences in behaviour of laboratory mice have been observed, particularly between albinos and wild-coloured or black ones. Albinos appear to be particularly sluggish in their reactions, with less hasty movements. They make hardly any attempt to escape, occupy themselves more intensively with new objects in their surroundings; their activity and rest phases alternate more uniformly; and their activity organization is less influenced by the 24 hour periodicity. Their fertility, however, also seems to be somewhat lower. Experiments on familiarization to new rooms carried out by Ernst Spangenberger in the author's research group showed that, compared with black mice, wild-coloured ones are more motile, interrupt their movements more frequently, jump more, and defecate and urinate more quickly. They seem to be more restless and take flight more easily on the whole. Mice with a *yellow dilution* allele turn out to be particularly active and restless, and tend to excessive reactions; animals with a pale colouring allele appear to be nervous and highly reactive, and their ability to survive when kept in crowded conditions is diminished. Other colour variants have stronger behavioural abnormalities with disturbances in motor co-ordination extending as far as stereotyped circling, trembling and cramps.

It is easy to obtain secure data on quantitative behavioural differences between different colour types with standardized experimental series on laboratory rodents, as they can be kept and bred in large numbers under the same environmental conditions. This is very much more difficult with large mammals under normal circumstances in zoos or wildlife reserves, where there are as a rule only very few really comparable individuals that can be used. Nevertheless, the author's research group has managed to collect some findings pointing in the same direction here as well. Albino dingoes that Lothar Schilling studied were, in comparison with tawny littermates, less timid in behaviour towards humans. The peaks of their diurnal activity shift strongly into the twilight of dawn and dusk, which seems to be a direct consequence of the lack of protective pigmentation in the red eyes of the albinos; this lack forces the animals to keep their eyelids almost closed in bright light and so makes them less confident under conditions of high illumination.

Preliminary data are available for sheep and goats. Quantitative records of the movements and social actions in a group of German heath sheep undertaken by Hedi Brentzel indicate that motility is somewhat higher in grey, dark-headed animals than in white ones. Corresponding studies by Sigrid Yavari-Kojabad on a flock of African dwarf goats showed highly intensity of motion and stronger social activity in dark wild-coloured and black animals than in white ones.

Eckehard Eich and Elisabeth Reichert observed a family group of Burchell zebras from East Africa, in which the otherwise rare allele that reduces that black colouring of the stripes to a light beige shade (so-called 'white' zebras) is present in a total frequency of about 50% (Fig. 12.2, p. 164). The 'white' animals distribute their activity more evenly over the course of the day than do the normal-coloured ones. The activity values of the 'whites' coincide with those of domestic horses and donkeys, while the values of the normal-coloured zebras differ distinctly from them. 'White' zebras are less shy of observers in the enclosure than the others. There are hints of certain differences in reactions to stimuli from the environment. A single albino donkey that was observed by Elisabeth Urban behaved similarly in comparison with grey, brown and black domestic donkeys. Its motor and social activities were lower, its daytime activity distribution more uniform and its trustfulness higher.

Elfie Bursch chose a different method to obtain preliminary measurements and observations of the reality of the sort of connections between colouring and behaviour that were part of the traditional lore of horse breeders. Instead of observations in a zoo or enclosure, or long, detailed study on very small groups, she gathered unique impressions of a large number of horses by testing 176 horses in stables with a simple three-level reaction scale. A person well known to the particular horse gave it a whack on the upper thigh, patted it on the neck, stroked it on the head and passed his hand over its flank against the lay of the hair. This sequence was repeated in the same order with each horse. The reactions were registered for each action in three categories numbered 0, 1 and 2 for 'no or hardly any reaction', 'laying back the ears and lightly defensive movement', and 'violent defence with snapping, rearing or kicking'. Finally, the sum of the values obtained from each animal was calculated giving a range of between 0 and 8. Since the results of this short sequence included all the chance elements of the horse's mood, its previous history, and its relationship to the experimenter, producing an extremely wide scatter in the values, at least the diversity of age, sex and breed had to be restricted so as to obtain statistically significant results. What was left over was sound evidence for horses from a narrowly limited area of origin only, for Hannoverian geldings aged over three years and for geldings aged over three years from all German light horse breeding regions, and only for the colour groups bay and chestnut. In both series, the chestnuts (reddish-brown without black manes and tails) reacted distinctly more than the bays (brown with black manes and tails). A further 'snapshot' study was then made of 273 horses of all breeds, sexes and ages that happened to take part in the procession on Rosenmontag (the last

Monday before Lent) in Mainz. This was carried out towards the end of the procession route as the horses were being ridden through a narrow, built-up street full of noisy onlookers, with bands playing, so that they were subject to an extraordinarily high stimulus load. Categorization of the way they happened to behave as they passed the observation point produced fundamentally the same results. While the bays generally behaved quietly and inconspicuously, the chestnuts were more frequently distinctly more vigilant in their behaviour, going as far as uneven movement and balking. The old lore concerning basic differences in behaviour between horses of these two colours is evidently more than just an old wives' tale.

There are many other individual clues for parallels between colouring and behavioural peculiarities for many other species of mammals but they have not yet been documented. In feral cats on the sub-Antarctic island of Marion, the *white-spotting* allele evidently has a certain influence on the development of the adrenal gland; no distinctive trait was found for the *non-agouti* allele and there are not yet systematic observations on the behaviour of the relevant animals. The black leopard has been repeatedly attributed a greater wildness (? restlessness) than the normal type, which has spots on a pale yellow background. A statistical evaluation of the litter sizes of zoo leopards seems to confirm this in terms of the stress concept, since the average number of young in the litters of black females is lower than that of normal-coloured ones. Another example taken from observations made by huntsmen can be cited for white and white-spotted roe deer, which seem more trusting and spring off later than brown ones when disturbed. Wolf Herre has reported that one can approach sleeping white domestic reindeer unnoticed without taking special precautions. They are supposed to be generally weaker than normal-coloured ones and can be approached unnoticed even when the rest of the herd is in a state of general excitement; Laplanders have the term 'tired colour' for such white reindeer. Black rabbits in an *agouti* versus *black* polymorphic population on the island of Skokholm (Wales) appear to be less timid than their wild-type conspecifics.

The linking of colouration with behaviour, and colouration with environmental appreciation, evidently takes place not only through quantitative shifts in the neurotransmitter system that supports information processing (shifts which are quite intelligible in terms of the biochemical connection between melanin and catecholamine syntheses) but also through changes at present more difficult to interpret in the information acquisition system (the sensory organs). That albinos suffer a loss of information in bright sunlight because the lack of protective pigments in the iris forces them to close their eyes more, as has been observed in rats and dingoes (Fig 2.5, p. 16), is a functionally necessary connection. This is also the case in non-albino, white reindeer. Albino rats, however, seem to react less to olfactory stimuli as well. Eye defects ranging up to blindness as well as deafness occur in dogs with the so-called English spotting. The combination of blue eyes and deafness in dominant white, non-albino cats, however, is a case of the linkage between two different loci being the origin of a relation between colouring and sensory organs. In the thorough work of Willys K. Silvers on the comparative genetics of the coat colours in mice, some different colour types combined with especially active and more highly reactive behaviour were found to involve rather far-reaching degenerations in the area of the inner ear.

Possible relations of pigmentation to hair structure should not be neglected in this connection. Clyde E. Keeler discovered that albinos in an Indian tribe in Panama had finer hair on their heads than the rest of the members of the tribe, who had normal skin and hair colour. The hair of 'white' zebras seems to be longer in places than that of normal-coloured ones. It is said that chestnut horses have finer hair than others. Reliable information could only be obtained in these cases by comparative hair measurements. Such changes in the hair might suggest changes in the tactile sensitivity of the skin which, in turn, would lead to altered reactivity. Longer coats also affect insulation – the loss or retention of body heat. Lowering the rate of metabolism in long-haired types at high external temperatures can doubtless be achieved by reductions in activity.

It is not possible at present to generalize concerning the behavioural effects of changes in coat colour from the wild or normal type except in the case of the *albino* locus, since there are comparable studies for the albinos of laboratory rats, laboratory mice, and the dog (dingo). These studies agree with the observations on a single albino donkey and reports on albinos of other species; data on the 'white' zebras, which are not full albinos, tend to point in the same direction. The common factors are an increase in the sluggishness of reactions and a reduced readiness to take flight; correspondingly, the distribution of activity and rest throughout the day seems to be somewhat more balanced.

In laboratory rats, *non-agouti* black is connected with less timidity and less aggression than the *agouti* wild colour allele. The *non-agouti* black fox also seems to display less readiness to flee. Similar reactions have been reported for the *non-agouti* rabbit. However, the conjectures and data available on the leopard at present seem to indicate the opposite. Whether there is indeed a contradiction here will have to be clarified by exact studies; there is also the question of whether it really is a matter of the same or a completely comparable gene in all these cases. Changes in coat colour due to decreases in eumelanins, i.e. alleles of the *black* locus leading to more yellowish, reddish or brownish shades than the corresponding wild colour, seem frequently to involve attenuation of activity and reactivity. This is so for instance in the amber fox (with the *non-agouti* allele and a *dilution* allele), in the pastel and palomino mink, in the mouse and in the goat. A corresponding behavioural change in the 'tame' direction, however, cannot by any means be attributed to all colourings that are fundamentally lighter than that of the wild animal. In fact, the opposite change towards higher activity and more nervous reactions occurs with, for example, the *silver* allele in the house rat, the two-factor combination leading to the light grey sapphire mink, or some lighter-colour types of the laboratory mouse.

In general, these findings show that the activity and reactivity level of a mammalian species on the scale from wild animal to domestic animal behaviour can be shifted in varying degrees and different directions depending on the change in coat colour from the wild type. So a third principle of domestication can be formulated as follows.

Some mutants with coat colours deviating from the corresponding wild colouring are especially amenable to domestication due to favourable correlations in the realm of behaviour.

Synopsis

The coat colour of a mammal is related to the basic level of its activity, its reaction intensity and its environmental appreciation. The reason for this is probably to be found in the fact that up to a certain stage the pigments that determine colour – the melanins – and the catecholamine group of neurotransmitters that are to a large extent the basis of the information-processing system share a common biochemical synthesis pathway. Selection of certain coat colours can produce a behavioural change with a corresponding change in the stress system either towards attenuated behaviour and increased tolerance or in the opposite direction. Combinations of the alleles of single colour genes that deviate from the corresponding wild-type increase or alter their effect on behaviour. It follows that the strategy of selecting and combining certain coat colour types can produce direct effects on domestication.

9 | Coat colour selection

As a rule, a coat colour deviating conspicuously from the norm in a wild animal puts it in danger, especially in a gregarious species, as has already been explained in Chapter 2. This danger arises because the animal stands out more in its environment, to which the normal colouring is adapted, and because of its oddness within its social unit, which makes it an object of attack from outside and of social aggression from within. On the other hand, an aberrant colour could also offer the animal a greater protection against predators. If a carnivore has learned from experience to search for an image fitting the norm of a particular species, its attack behaviour is initially inhibited by a conspicuously different, exceptional appearance that does not match the preconceived pattern. The attack may be carried out only hesitatingly, which gives the prey a better chance of escaping than its normal conspecifics, which are subject to uninhibited pursuit immediately upon recognition. Laboratory experiments with ferrets, offering variously coloured rats as prey, and field experiments with Old World kestrels and wild-coloured and white mice as the choice of prey, demonstrate this both for carnivorous mammals and for birds of prey.

An animal that hunts with its sense of vision and has learnt searching images for prey is not completely neutral when confronted with a conspicuously deviant individual. The predator either reacts somewhat hesitatingly or can make the deviant a special target as it is easier to keep an eye on. Neither does a human hunter act neutrally in the same circumstances. His reaction to an unusual phenomenon in a familiar context may be determined by responses ranging from a desire to possess it to fear of the unnatural, which, in the context of many religions, means the supernatural. Human attempts to integrate all the phenomena of the environment into a uniform conception of the world with as few problems as possible, and so to have some sort of explanation for special cases, suggest a transposition of such rare exceptional cases into the domain of cult, at least among peoples with an animistic religious world view. This might form a basis for the association of the sudden appearance of a conspicuously different animal with ideas of supernatural signs, or interpretations of good or bad fortune. If such an attitude arises it lays the foundations for selective hunting or not hunting according to coat colour, for selective live capture, keeping and breeding not directed towards the usual use of the animal as a producer of food or raw material but rather determined by the cult value. That this idea of coat colour selection on cult principles really does hold generally can be demonstrated with many examples from the most diverse cultures.

The cult herds of the llama and alpaca, the domestic animals of the Incas, were kept separated according to colour. Each deity was attributed a favourite colour. One hundred brown animals sacrificed from August to September were supposed to protect the new maize fields from inclement weather. The sacrifice of 100 white llamas in October was linked with ideas of bringing rain. Sacrifice of 100 white animals was also supposed to influence the sun to shine and so promote growth in the fields. Finally, the maize harvest in May had to be favourably influenced by the sacrifice of 200 animals of all colours. A main symbol of the power of the Inca rulers was the napa, a white llama (Fig. 9.1). At the main temple in Cuzco, a white llama was sacrificed every morning. White alpacas served as offerings for the sun god, black ones for his son. Before the period of Inca rule, a tribe at Lake Titicaca even worshipped the white alpaca as the principal deity.

The prairie Indians of North America regarded the coat of a white bison (buffalo) as sacred and it was revered as a talisman in the hunting cult. On the Samoan Islands in Polynesia, a bush spirit could appear in the form of a white dog whose behaviour was considered to be a good or bad omen. So-called 'white' elephants were reserved for the king as ceremonial animals in Siam (Thailand). Albino Bali cattle come mostly from sacred herds. Finally, there is a correspondence between the sacredness of white cattle in the Hindu faith and in ancient European civilizations.

For instance, white sacrificial cattle played a role for the Romans, the Celts and other peoples in Antiquity. In the Celtic religion, black cattle were a symbol of death, brown ones of fertility and white ones of the sun

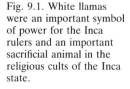

Fig. 9.1. White llamas were an important symbol of power for the Inca rulers and an important sacrificial animal in the religious cults of the Inca state.

cult. A sacrifice of 300 white cows with red ears is recorded in the traditions of pre-Christian Ireland. The price of such white cows was 18 times that of the usual-coloured ones. Cattle of this colour type also played an outstandng role in the tradition of Wales. This white with red ears is exactly the colour combination still found today in the relict Chillingham Park cattle in Northumberland, which may therefore be the last surviving type of pre-Christian sacred cattle in Great Britain (Fig. 9.2). A white mare is supposed to have played an essential ritual role in the election of Celtic kings. Palomino horses were kept for use in processions in pre-Christian Norway. Black animals were usually selected when dogs were sacrificed to the goddess Hecate in Ancient Greece. Finally, a colour symbolism associated white with purity and innocence, black with evil and hell. Black cats were used in devil worship in Medieval England and were often connected with witchcraft in the Middle Ages in Europe. The 'black sheep' became a proverbial name for someone deviating negatively from the norm for his group, for example from his family. Killing a white hart used to be considered bad luck. Folk medicine also included similar ideas; for example, a piece of cloth that had been tied around a black goat was supposed to help against mumps.

Selection on the basis of colour, however, did not have a background only in cult worship: practical notions were sometimes also certainly decisive. For instance, the Roman author Columella reports that shepherds preferred white sheepdogs as they could not be mistaken for a wolf raiding the herd at night and so be attacked and killed instead of the wolf. This ancient colour description in fact matches many modern breeds of sheepdog, for example the gigantic Pyrenean mountain dog, the large Maremmano–Abruzzese breed of the Italian mountains, the large to gigantic Hungarian breeds komondor and kuvasz, the equally large Tatra dog, or the more medium-sized sheepdog breeds in Central to East Europe. Finally, it was not cult interest but pure curiosity which led

Fig. 9.2. White cattle played an important role as sacrificial animals amongst many peoples in the ancient world. Feral cattle in Chillingham Park, Northumberland, shown here, may be relicts of the sacred Celtic cattle type.

to the selection involved in the domestication of the laboratory rat and golden hamster as mentioned in Chapter 3.

The various cult ideas connected with coat colours deviating from the wild colour in domestic animals and also in important wild game animals – particularly with shades of black and white – suggest that corresponding selection may well have taken place in the early phase of domestication. Such ideas may more than once have supplied the impetus for early primary domestications. Some information on the frequency of unusual colours in the first phase of the history of individual species of domestic animals can be obtained from analyses of early works of art such as have been undertaken by Burchard Brentjes. Early written records such as epics and sacred writings also provide important data. For instance in the story of Jacob in the Old Testament Book of Genesis, the historical kernel of which is to be sought in the second millennium BC, spotted, coloured and black sheep (Fig. 9.20) and coloured and spotted goats are mentioned. The existence of the *white spotting* allele for cattle is on record from at least the sixth millennium BC and the white colouring from the third millennium at the latest. Pale and spotted coat colourings were to be found in the dog from the third millennium BC at the latest also. The existence of different colourings in the horse is evident from the second millennium onwards.

Estimating the distribution and the significance of coat colours deviating from the wild type in domestic animals may be facilitated by a short review of the various species. As mentioned at the end of Chapter 2, the dingo, which obviously returned to the wild while still in the initial stage of domestication, lacks the allele responsible for the grey wild colour, i.e. the wolf colour (Figs. 1.4 (p. 3) and 2.5 (p. 16)). Apparently this was also the case for other primitive dogs in Africa, South Asia and Polynesia (Figs. 9.3 and 9.4). That allele was probably introduced into the populations in some places secondarily by the breeds imported from Europe. In effect, this is so for all primitive dogs in countries where there were originally no wolves. The loss of this colour allele in dogs must therefore have occurred before the dog spread to the regions in question and so was a result of the primary domestication itself. The occurrence of the wolf

Figs. 9.3 and 9.4. Madagascan dogs are examples of primitive dogs in tropical lands free from wolves. Yellowish colours (dingo colour) and comparatively large white spotting are the most frequent colour types; the wolf colour seems to be lacking where European breeds have not been imported.

colour in dogs from regions inhabited by wolves, especially from the far north (Fig. 2.25, p. 33), can be understood in terms of secondary interbreeding, for which there is some evidence. Wolf-grey, however, also is clearly uncommon in the European improved breeds, in comparison with the various other colours. It is frequent in the various watchdog breeds and in dogs from northern countries. Primitive dogs from tropical lands have mainly reddish-brown and reddish-yellow coats in various shades, black or white, and white spotting on reddish-brown or black backgrounds (Figs. 9.3, 9.4, 1.4 (p. 3) and 2.5 (p. 16)). The yellowish-reddish colouring predominating in primitive dogs can be encountered in some wolf populations varying from merely hinted at to very marked; so too can both black and white. In principle, then, it is quite conceivable that this is the way the initial selection for domestication from the wolf to the dog began by eliminating the allele that determines the typical wolf-grey right from the start.

Certain studies on the distribution of coat colours in modern cat populations make it possible to draw up world maps of the allele frequencies for all important colour genes. These maps show the focal areas with especially high frequencies of particular alleles and adjacent zones of gradual decrease in frequency. For example, the *non-agouti* allele responsible for the occurrence of black cats occurs with a frequency of over 80% in England and the coastal region of northwest Africa (Morocco, Algeria). Frequencies of between 70% and 80% are found for this allele in southern Scandinavia, the French and Iberian coastal strips, and along the old trading routes from the western Mediterranean through the valleys of the Rhône and Seine rivers to the French coast of the Channel. Then there are frequencies of between 60% and 80% in all areas bordering on the central and eastern Mediterranean (Fig. 9.5). Such a distribution type clearly concentrated in coastal countries and shipping centres suggests that this allele had certain advantages for cats spreading as ship's cat populations. It is an open question whether these advantages were a higher tolerance of black cats to the special environment of the ships of past centuries or a special preference on the part of seafarers. By contrast, the distribution map of the blotched-tabby pattern in comparison with the (wild type) striped tabby pattern shows a picture of a distinct centre of high frequency, with a gradual decline with distance, and no correlation to coastal areas. This centre is in England, where over 80% of the cats have this allele. The proportion is still above 70% in parts of eastern Ireland, southern Scotland, and northwestern France. In the rest of the British Isles and the greater part of France proceeding from the northwest it is between 60% and 70%. Further decreases in frequency occur going from the French region towards the Iberian Peninsula on the one hand and towards Central Europe and the Apennine Peninsula on the other. Finally, in eastern Europe and the southern and eastern Mediterranean countries, the proportion of the

Fig. 9.5. Map of the frequency of the *non-agouti* allele in European domestic cat populations. (Redrawn from Todd, N.B. (1977), Cats and commerce. *Scientific American*, **237**, 100–7.)

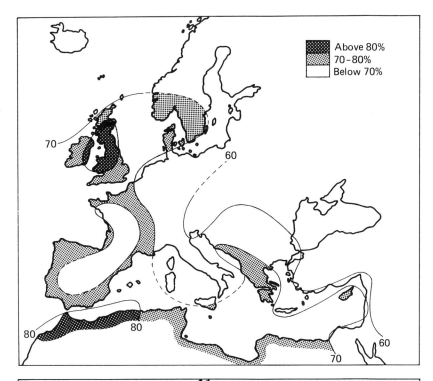

Fig. 9.6. Map of the frequency of the *blotched-tabby* allele in European domestic cat populations. (Redrawn from Todd, N.B. (1977), Cats and commerce. *Scientific American*, **237**, 100–7.)

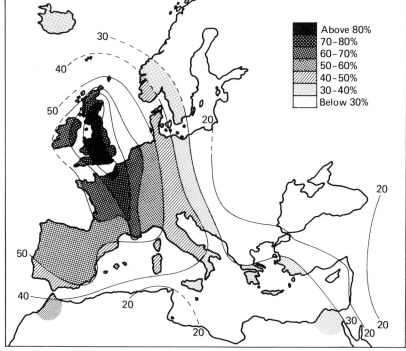

blotched-tabby pattern amounts to only 40% at the most and in some places drops below 20% (Fig. 9.6). An analysis by Neil B. Todd suggests that the extraordinary accumulation of this allele in the British Isles came about in the last 300 years. Since the seventeenth century, British cats have been exported into countries that were still without cats, for instance at various periods to the New England states in North America, to Canada, to Australia and to New Zealand. On the assumption that the present allele frequency in each population newly created in this way still reflects the allele frequencies in the country of origin, i.e. the British Isles in this case, at the time of their export, the close connection between the date of export and the frequency of the alleles in the area of origin implies that the allele occurred with a frequency of 40 to 50% in England in the seventeenth century, 50 to 60% in the eighteenth century and 70 to 80% in the nineteenth century. There are centres for other cat colour alleles outside Europe and the Mediterranean region. For instance, the type of *white spotting* that strongly reduces the remains of the original coat colour in favour of white areas occurs very frequently in South Asia, from where it radiates to countries colonized from there. For example, there is a comparatively high percentage of this white spotting in cats on Madagascar, where the culture and population derive largely from south Asia (Fig. 9.7). In most regions, the pure wild-type colour makes up only a small portion of the colour variation of the domestic cat (Fig. 9.8).

Albino ferrets are more frequently encountered than those with wild colouration. Hunters usually prefer them, as they are easier to handle when hunting. The recently farmed predators, fox and mink, are largely selectively bred for fashion reasons for colours deviating from the wild type.

Among horses, the two known wild-colour types – namely the variations of reddish yellow-brown of the Przewalski horse and the

Fig. 9.7. Spotting with an extremely high proportion of white in a Madagascan domestic cat.

mouse-grey of the extinct East European wild horse (tarpan) – are only rarely found in domestic horses. The domestic colours in various dun, brown and black shades, grey, and piebald clearly dominate and are found exclusively in improved breeds (Figs. 9.9 and 9.10). Only in a few very primitive local breeds, mainly in East Europe and Scandinavia, is the mouse-grey tarpan colour still represented. It is typical of the back-crossing of the Polish konik in the tarpan direction (Fig. 3.19, p. 53). Unlike in the horse, the wild colour is very frequent in the donkey and is the normal colour of the domestic animal in many populations (Fig. 3.21). Nevertheless, in special, improved breeds it has been noticeably

Fig. 9.8. Wild-type spotted and striped patterns in the domestic cat (cf. Figs. 3.12 and 3.13) compared with the blotched-tabby pattern on a wild-coloured (centre) and a reddish coat (right) (the differences between the degree of patterning are independent of the basic coat colour).

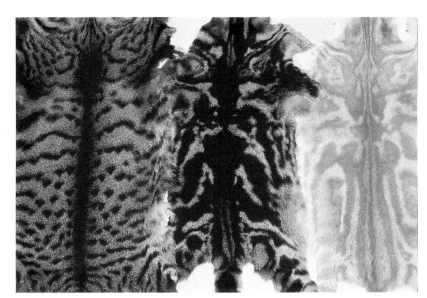

Fig. 9.9. White colouring ('grey') in the horse.

repressed in favour of other coat colours, for example white in the large Egyptian–Arabian donkey and the dark brown to black of the huge Poitou donkey and the Catalanian donkey of Spain (Figs. 9.11 and 9.12).

Improved breeds of pig also lack the wild colouring completely. In Europe and regions stocked with animals from there, white is the main colour in pigs used for meat production today. It originated chiefly in the Danish and German landrace and the English Yorkshire pigs (Figs. 1.5 (p. 4), 10.1–10.5 (pp. 149–51)), who obtained it in their turn via the earliest introduction of Asian breeds. The black and white belted colour is probably of European origin (Fig. 9.13). Primitive local breeds all over the world are often spotted with black, grey, yellow or red. There are

Fig. 9.10. Piebald colouring in the horse.

Fig. 9.11. Uniformly dark-brown colouring in the donkey (Poitou breed).

Fig. 9.12. White colouring in the donkey, with a shoulder cross still just recognizable.

Fig. 9.13. Belted colouring in a pig (German hybrid hog).

Fig. 9.14. Black Polynesian pig.

pure-black pigs mainly in East Asian improved breeds, in South Europe and in Africa. The wild colour is typical of the primitive group of Papua pigs, though black ones occur among them also. These are evidently *non-agouti* black, as the piglet striping of these black pigs, typical of wild animals, is visible only when the light strikes them from a certain direction – as is also the case for other patterned mammals (Fig. 9.14).

In the two domestic camels, the Bactrian and the dromedary, the wild colour is the rule, although deviating light and dark colours, and spotting, are also found, particularly in the dromedary (Figs. 9.15 and 9.16). The guanaco wild colour is extremely rare in the llama, which is predominantly variegated in different ways in white, brown and black (Figs. 3.27

Fig. 9.15. White Bactrian camel with its young and a wild coloured animal (right).

Fig. 9.16. A light and a dark-brown dromedary.

(p. 61), 9.1, 9.17 and 9.18). No wild colour at all, be it that of the guanaco or that of the vicuna occurs in the alpaca (Figs. 3.29 (p. 62) and 9.18), which is normally coloured white, dark brown, black and grey in several shades, but rarely spotted.

In addition to the wild colour, there is a palette of other colours such as black, white, reddish, white-spotted and silver in the domestic reindeer. In Alaska, black ones are preferred for breeding, since they are considered to be the best. Steel-greys are also supposed to be strong animals, spotted ones less so and whites hardly at all. The colour range in domestic cattle is rich. Many colours are directly characteristic of single breeds or groups of breeds and have even supplied breed names. Zebus are

Fig. 9.17. A section of the colour palette of the llama.

Fig. 9.18. Weak (left foreground) and complete (right background) expression of the vicuna wild colour in the alpaca (brown animals at the back and in the middle), which came about by interbreeding vicuna with an alpaca group. In purebred alpacas this colour does not occur.

predominantly shades of brown, grey, black, white and white-spotted (Figs. 2.8 (p. 19) and 3.33 (p. 66)). The wild colour was different for aurochs bulls and cows in northern Europe – blackish for the former and brownish for the latter (Fig. 3.31 (p. 65)) – although in the Mediterranean region the aurochs bulls were evidently also brown. Now the wild colour is found only among a few local cattle breeds in the Mediterranean area. There is a similar diversity of colours for the yak; about 60% are white-spotted (Fig. 3.34 (p. 67)). Bali cattle are mainly the same colour as the banteng, i.e. black bulls but reddish brown cows (Fig. 3.32 (p. 65)) but there are also black, white, yellowish, grey, and white-spotted animals. The same is so for the mithan and the domestic water buffalo.

Like domestic cattle, domestic sheep seldom display the wild colour. However, the Soay sheep, which has lived ferally since prehistoric times and is at an early level of domestication, still embodies the wild colour (Fig. 3.37, p. 70). Less primitive local breeds and all improved breeds have colours differing from the wild type. The following are widespread: white, various shades of grey, black, brown in shades from copper to blackish brown, and various spottings ranging to breeds in which only the head is different from the body colour (e.g. black, brown and white-headed sheep) (Fig. 9.20). In the domestic goat, the wild colouring with a black dorsal stripe, shoulder cross, belly stripe and leg markings in various shades of the brownish basic colour is encountered only in the so-called 'coloured' goats (Fig. 9.21). Particularly frequent colours are black and white, brown and various kinds of spottings such as the extreme of the half-white/half-black body of the Valais goat (Fig. 9.22).

The wild colouration is also much less frequent than others in small

Fig. 9.19. Black Forest cattle, example of one of the white-spotted colour types of cattle.

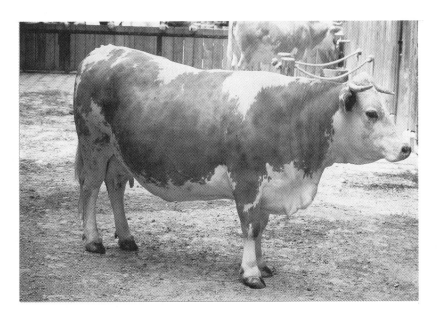

domestic rodents. Albino white is the most widespread in the rat and mouse (Figs. 3.45 and 3.46 (p. 76)). The laboratory mouse has been the subject of intensive studies on coat colour heredity in mammals precisely because of the huge number of colours found in it. The wild colour is not found at all in the golden hamster, which was derived from a golden mutant of the wild form in the first place, but this is made up for by many other shades (Fig. 3.47, p. 77). The wild colour is scarcely found in the guinea pig, the dominant colours being black, brown, white and white spotting (Figs. 2.13 and 2.14 (p. 23), 3.48, p. 77).

The rabbit has also become a classical subject for the study of coat colour heredity mainly due to the work of Hans Nachtsheim. Since about the end of the Middle Ages, one can follow the increasing displacement of

Fig. 9.20. The colourful picture of a local breed of sheep: white, black and variously spotted animals together in one flock (Attica, Greece).

Fig. 9.21. Wild colour type with some white spotting in a billygoat in bushland in southwest Madagascar.

the wild colour by more and more new variants (cf. Chapter 2). In the sixteenth century there were, in addition to the wild colour, yellowish, brown, and blue dilution variants from it as well as albino white (Fig. 9.23), black and 'Dutch spotting'. The *silver* allele was added in the seventeenth century; in the eighteenth and nineteenth centuries appeared also the so-called Himalayan colour from the albino series, the 'English spotting' (largely white with small, pigmented spots), the black and tan colour from the allele series of the *agouti* locus and two alleles of the *black* locus, the harlequin and steel-grey. In the present century, the main additions have been further alleles of the albino series, the various light to dark chinchilla colourings as well as a *white* allele, and numerous

Fig. 9.22. Section from the colour palette of the domestic goat (African dwarf goats).

Fig. 9.23. Albino domestic rabbit. Albino white was one of the earliest deviations from the wild colour type in rabbit breeding.

combined types of the various colour genes have been bred, some of which are typical for the appearance of various breeds.

This review makes it clear that the wild colour of domestic animals is mainly confined to unimproved breeds or to very primitive, local breeds. It has disappeared altogether nearly everywhere in special improved breeds. So selection for colour seems to be an essential process in the course of domestication which may even have set off the process of domestication in the first place in some species, and attained increasing importance in breed improvement in others. The impetus for this is often to be found in cult interests rather than in practical considerations. Interest in the unusual, the conspicuous, the deviation from the norm, still contributes to the preservation of some local breeds of various domestic animals that have hardly any practical use any longer; this was also the direct motive for domesticating the rat and golden hamster. The connection between coat colour and behaviour shown in the previous chapter, and increased human willingness to keep an animal of a deviating colour, creates an ideal combination for domestication. In the long run, it led to the creation of domestic animals without their being used for any directly practical purpose initially. So animals that were coloured conspicuously differently, whose deviation was the point of departure for their special treatment, possessed a significance of the first order in domestication.

Synopsis

The selection of coat colour variants distinctly deviating from the wild norm seems to be largely connected with human interest in the unusual. This readily assigns such particularly conspicuous animals a role in the cults of prehistoric and early civilizations. So the motive for domestication is often to be sought in cultural interests and the desire to acquire something unusual; no immediate, purely practical purpose in everyday life need have been involved. At the same time, the special amenability of many deviantly coloured animals for domestication set animal-keeping, which was begun independently of any secular considerations, on the road to success. In the same way, such a method may have accelerated the introduction of improved breeding of animals from the mass of still hardly altered primitive domestic animals. Selection for colour seems to be a central factor in domestication generally.

10 | Limits of endurance

The decline in environmental appreciation that places the domestic animal at a lower stress level than the wild one under the same environmental conditions involves a drop in its adaptability to acute stress. The reduced size of the adrenal gland is accompanied by decreased corticosteroid production. Experiments imposing stimulus load on domestic foxes (see Chapter 12) showed that the secretion of corticosteroids was comparatively higher than in the control group of animals though it did not reach their absolute levels as a rule. As is the case with other organs, lack of constant exercise evidently impairs the efficiency of the stress organ, the adrenal gland, in situations of acute stimulus load. For instance, laboratory rats seem to have more difficulty than wild rats in adapting physiologically to greater cold in winter. While the latter can maintain their body weight, the former lose weight. Such kinds of load, or fights and exhaustion, have no measurable effects on the ascorbic acid (vitamin C) content of the adrenal glands in wild rats, although they lead to very large decreases in ascorbic acid in laboratory rats. Due to the connection between coat colour and behaviour, albino mice are evidently less resistant to cold than wild-coloured laboratory mice. Improved breeds of meat pigs with particularly small adrenal glands have very little adaptability to physiological loads of various kinds.

The causal relationship between the decline in environmental appreciation and an increase in susceptibility to extreme environmental conditions leads ultimately to increasing restriction of the survival capability of the domestic animal to a rather narrow domain under human supervision and control. This is more so as the ability to acquire and process information is limited in the direction of improved breeds.

Alterations in environmental appreciation and thus alterations in the endurance of an animal can be attained not only by breeding measures but also by the way the growing individual is kept. Further insight is provided into this by experiments conducted on laboratory rats by David Krech, Mark R. Rosenzweig and Edward L. Bennett. Rats weaned at an age of three to four weeks and placed alone in cages without visual contact with others for a month, i.e. that were reared in an impoverished environment, performed worse in problem-solving tests than did control rats which, in this period, lived in large communal cages with varying devices to encourage exploratory behaviour. In addition, the control rats were placed each day in a test apparatus with increased exploratory possibilities. These rats from an enriched environment had cerebral cortexes about 3% larger (cf. Chapter 6) with about 4% more activity of the enzyme cholinesterase, which breaks down the neurotransmitter acetylcholine, in the same section of the brain (cf. chapter 7). Thus, withholding opportunities for social contact and impoverishment of the explorable, inanimate environment have the effect of reducing the capability and activity of an animal's information processing. The animal is therefore at a lower stress level under comparable external conditions.

Modern, rationalized methods of animal breeding make use of this principle by creating husbandry conditions that impoverish the environment. Piglets and calves are removed from their mothers early on. Although they are then raised in confined conditions together with others of the same age, the opportunities for social learning are reduced. This is achieved by limiting spatial and motor development and preventing encounters between different age levels such as are provided within a natural social structure. Continuous crowding of animals is evidently facilitated by a high degree of uniformity in the environment of the stable. Current figures from modern high productivity breeding of pigs clearly show the narrow confinement and social crowding that are possible wihout preventing successful growth and later successful reproduction in improved breeds of domestic animals. In the Federal Republic of Germany, for example, the guideline for communal keeping of eight to ten piglets each weighing up to 20–25 kg is set at an area of only 2.5 m², i.e. no more than 0.25 m² per piglet. The minimum area required for animals weighing 30 kg is fixed at about 0.33 m² per piglet and at about 0.8 m² for gilts weighing 100 kg. The space provision for keeping pregnant sows in groups is 2 m² per animal, while a stock boar is supposed to have about 6 m² when kept alone. The standard measurements for a pen for one pregnant sow are a maximum width of 65 cm, a lying area of 190 cm long, and a total length including the feeding place and rear space not exceeding 250 cm. It is quite possible to achieve agriculturally advantageous results under such conditions. The breeding statistics for the Federal Republic of Germany in 1976 show that for the German landrace, the predominant breed, the average litter was 10.5 piglets per sow and the average number of piglets reared per sow per year was 19.11 (Figs. 10.1–10.5).

Such confined conditions, which could scarcely be reduced further due to the size of the animal alone, reach the limits of the animals' endurance as set by the genetic foundations of the improved breeds and acquired from their being kept in special conditions. Observations carried out by Erich Peter from the author's research group disclosed that, when sows are moved from boxes of 1.5 m² to single pens of 4.4 m² or vice versa, the relations between the proportions of various types of activity and the white blood count change in ways that could be interpreted as signs of increased stress when the animals were kept in close confinement. The motor activity in the small pen decreases sharply; the proportion of eosinophil granulocytes related to the adrenal cortex system drops, i.e. the sows react physiologically as though subject to greater stimulus load. The reason why they still attain better reproductive results and more uniform litters with a somewhat higher number of live-born piglets under such conditions may be sought in the reduction in psychosocial stress when the sows are kept alone. This evidently more than compensates for the stimulus load created when they are closely confined (Fig. 10.6).

Climate, feeding and other external stimuli are largely standardized

Figs. 10.1 and 10.2. Reduction of the space required by an individual for intensive pig production: keeping young pigs in cages from the age of 22 days until reaching a weight of 20–25 kg (two to three months of age). Radiators maintain the necessary temperature.

and kept constant in modern high productivity husbandry. The result is that, although no decisive role is played by the probably extremely high psychosocial stress resulting from the crowded conditions, special stimulus complexes, to which the animal has had hardly any chance of becoming accustomed, can lead to fatal, acute stress reactions. This is

Fig. 10.3. The minimum space requirement set for high production pig breeding scarcely exceeds the room taken up by the animal's body. When kept in groups, the animals are crowded close together (fattening pen with partially slatted floor).

Fig. 10.4. Pregnant sows kept in single pens. The animals remain here until the third day before farrowing.

because the organism, in particular its adrenal gland, is completely unprepared for them. So, as the rationalization of pig-keeping increased and improved breeding for meat production progressed, the death rate in pigs during transport went up considerably. Losses during transport to

Fig. 10.5. Pen for piglets where they are kept with the mother until the age of 22 days before she is removed to breed again (see also Figs. 10.1 and 10.2). The sows are brought here three days before the calculated farrowing date.

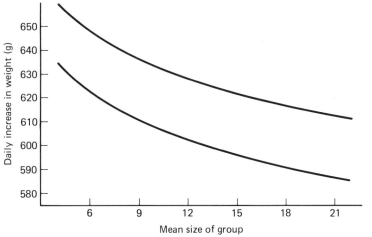

Fig. 10.6. Diagram of the effect of psychosocial stress in modern pig fattening: mean daily increase in weight versus the number of fattening pigs kept in the same sty (group size, cf. fig. 10.3). Upper curve: Danish-type stabling with a solid floor in which the lying space and feeding place is partially separated from the defecation area by a dividing wall. Lower curve: stabling on slatted floor constructed of parallel planks separated by narrow gaps. The lying space and defecation area thus become one unit; urine and faeces can drop through the slits at all places and be collected on a lower level. The different positions of the two curves makes it clear that, in addition to psychosocial stress, stress due to the different construction of the housing environment plays a considerable role in the pigs' maturation. (Curves plotted from values in Blendl, H.M. (1977), Haltungssysteme für Mastschweine. In *Schaumann Handbuch der tierischen Veredlung*, ed. E.C. Kamlage, Osnabrück.)

slaughterhouses amounted to DM76.6 million in the Federal Republic of Germany in 1976. The animals, accustomed to being handled uniformly in the sties, to uniform temperatures and changes in illumination as well as to certain social partners from the same husbandry group, are greatly overtaxed by the sudden high excitement. They are driven together on to the loading area under completely different external conditions. The ramps and floors are sometimes very slippery and the pigs are forced together with little known or quite unfamiliar animals in very close, sometimes even too close, quarters. Finally there is congestion and heat in the transport vehicles, which often have insufficient ventilation and temperature regulation. The consequences are stress reactions, even sometimes death from stress. The special preconditions create the apparently paradoxical situation that domestic animals bred for increased psychogenic tolerance display stress reactions like those of highly susceptible wild animals. The lowering of the adaptability to acute stimulus loads involved in the decline in environmental appreciation has evidently reached its limit.

The limits of endurance reached by domestic animals in the course of improved breeding are very different for each species, and for each breed within a species. The guidelines for modern standardization of breeding and husbandry conditions in the Federal Republic of Germany reflect the different uses for the different species well (Fig. 10.7). Stanchion stalls for the fattening of young bulls provide for a standing length of 120–130 cm and a standing width of 70 to 80 cm for animals up to a weight of 300 kg, a length of 140–160 cm and width of 90–100 cm for weights up to 600 kg, i.e. a situation fully comparable to that for the close keeping of pigs (Fig. 10.8). The situation for keeping sheep in pens all the year round is similar. Here the guidelines lay down a required area of about 1 m² per ewe without lambs and about 1.5 m² for ewes with lambs, between 0.5 and

Fig. 10.7. Calf-raising in group penning. The animals are brought here at about 12 days of age.

0.75 m² for a fattening lamb and 1.5– 2 m² per ram in group keeping or 3–4 m² when kept singly. These three types of animal bred for meat contrast with horses. The minimum area for a stall for a brood mare is set at 12–16 m², and 10 m² for a frequently exercised saddle horse. In addition, a run connected to the stable or, even better, regular grazing in the field is recommended. The complete difference in use is reflected in the much higher spatial requirements compared with the meat-producing animals (Fig. 10.9).

The limitation of the survival capacity of improved breeds to an environment created and kept as constant as possible by man is in many cases based on a number of factors besides decline in environmental appreciation, with decreases in the adaptability to acute stimulus loads. Where selection has occurred to eliminate the original coat, new require-

Fig. 10.8. Fattening young bulls in stanchion stalls, a spatial situation in modern beef production comparable to keeping pigs in confined areas.

Fig. 10.9. Cows are allowed greater freedom of movement when kept for milk production. Although each animal has only limited space in its lying box, it can move about within the whole building and is put out into the pasture by day in summer.

ments of minimum external temperature level and uniformity have arisen. If the temperature falls below a minimum threshold for a long period, the regulatory capability of the organism is overtaxed because the insulation originally provided by the coat is missing; this eventually leads to collapse due to cold. Temperature regulation is not only important for hairless dogs and cats but also a reason for the necessity of maintaining certain temperature zones in modern pig sties that lack any insulating material. In the extreme case of rearing piglets on a grating floor without bedding, a minimum temperature of about 26 °C is reached (Fig. 10.1) The other extreme, which needs more work but uses insulating material, namely piglets kept on solid floors strewn with bedding, allows temperatures to be lower by 4 to 6 deg.C in each case. A further important restriction on capacity for independent survival applies to breeds of domestic animals in which disturbances to parturition occur to a high degree. For instance, a high percentage of Persian cats, one of the extremes of improved breeds in domestic cats, have to depend on human help when giving birth, if the young are to survive. The behavioural deficiencies begin with disturbances in the course of labour, then the insufficient removal of the placenta or failure to remove it at all from the newborn which are threatened with suffocation, the frequent failure to sever the umbilical cord, the insufficient drying of the young, and finally the failure to eat the afterbirth. This completely passive behaviour of the mother cat endangers the survival of the newborn at almost every point unless the breeder intervenes to help.

Synopsis

The decline in environmental appreciation of domestic animals is accompanied by a lowering in their adaptability to acute stimulus loads due to functional alterations in the adrenal gland. This innate and breed-specific situation can be further intensified by special rearing methods in stimulus-impoverished environments. In this way, the spatial require-ment for an individual animal may be lowered to close to the minimum set by the physical size in rationalized keeping of breeds improved for meat production. However, the limit of susceptibility, of succumbing to unfamiliar loads, is reached at the same time. Special directions of selection for various utilization purposes have resulted in improved breeds of domestic animals whose capability of independent survival outside of the largely standardized environmental conditions provided for them is severely limited in many cases.

11 | Taming and return to the wild

Fundamentally, taming is not equivalent to domestication. Neither is it a necessary condition of domestication even though domestication processes may have occurred mostly with more or less tamed animals. Likewise taming is not a first step towards domestication. Both states, tamed and untamed, are possible in domestic as well as wild animals, although it is true that domestic animals are usually tamed and wild animals usually untamed. Just as there are wild animals kept in captivity that are completely tame, so there are feral animals that live untamed in the wild. The difference is merely quantitative: because of the lower significance of environmental stimuli, i.e. because of a reduced environmental appreciation, a domestic animal seems basically tamer than a wild one insofar as it maintains shorter flight distances.

Taming is a condition for dealing with domestic animals in a confident, unproblematic way. The process of taming involves learning by the animal, during the course of which the flight distance and possible aggression towards humans is gradually reduced. A fully tamed animal finally no longer associates humans with danger. Since domestic animals mostly go through this process of learning by habituation very early in their lives, even before any visible avoidance behaviour has been developed, the animal's keeper as a rule is not aware of it. So the domestic animal seems to have been tame from its birth onwards even though this is in fact not the case, beyond the fundamental behavioural traits of the domestic animal.

On reaching sexual maturity, many male domestic animals again become difficult to deal with, despite or even just because of their tameness. Humans increasingly become the target of aggression, since taming has either eliminated the animal's fear of them or even prevented its development altogether. To overcome this problem, the animals are sometimes castrated, which, in this connection, is a kind of 'physiological taming'. In place of aggressive, obstinate animals that are difficult to manage, it creates ones with friendly behaviour that are easy to handle, i.e. that are on the whole tame.

Castration neutralizes the activity of the gonads and leads to the elimination of the male sex hormone testosterone that is responsible for typical 'male' behaviour. Particularly in male domestic animals that, in contrast to their wild ancestors, are in season more or less the whole year round (cf. Chapter 4), this has a considerable physiological taming effect. General 'domesticity' is furthered in this way.

There are differences in the stress system of male and female mammals. Only corticosteroids in the blood that are not bound to globulins are biologically active and so participate in stress events. This binding capacity is lower in males than in females and at the same time the domestication effect seems to be less in males. For instance, the difference in corticosteroid levels in females was higher than in males in an experiment conducted by D.K. Belyaev and L.N. Trut on silver foxes that had been subjected to selection for domestication and ones that had not. Male animals seem to have a lower level of physiological endurance than females. This is the point where castration intervenes.

Because of interactions between sex hormones and growth hormones, alterations occur in juvenile growth after early castration. Although castrating cats early leads to increases in bodily growth, it apparently reduces the brain size. There is evidence that castrating humans may have the same effect. However, it is not yet known how the information-processing system is affected. Castration probably affects the stress process at several levels. It helps to make the animal manageable, even when the selection that has taken place during domestication and the additional learning processes during taming have failed in this respect.

Even early stages of domestication often make practical use of castration. For instance, reindeer-keepers in the far north mostly allow only young stags to reproduce and castrate older ones to prevent the development of old, aggressive stags. The production of oxen, animals that can be put to work without any problems, is an age-old custom in cattle husbandry and is still practised where cattle are used as work animals. Gelding makes horses more manageable as saddle animals. Castration prevents the nuisance caused by the smell of the urine-marking of the adult tom-cat. And there are further examples of the positive effect for day-to-day handling of domestic animals for other species.

The opposite of taming is the return of the domestic animal to the wild, which, however, is seldom an option for improved breeds because of the limitations to which they are subject (cf. Chapter 10). Populations of feral animals make it possible to study the behaviour and population dynamics of domestic animals that have not been under human influence and so not tamed, and to draw direct comparisons with the situation of free-living wild animals. Where the return to the wild occurred a long time ago, it also provides the opportunity of observing the effects of natural selection on populations of domestic animals. The only useful populations for such studies are ones that became feral in regions not inhabited by their ancestral species, so that no subsequent interbreeding with wild animals can confuse the picture.

These conditions are to be found mainly on some islands where the original stock of domestic animals were set free by sailors to form the basis for supplies of fresh meat on later landings. However, precisely at such places there are usually neither food rivals nor predatory enemies; so there is no selection pressure which would give rise to a retrospective evolution away from domestic traits in the direction of the wild type.

Comparative studies of the information-processing system and range of colour variability (of special interest here in view of the relationship between colour and behaviour), and observations of behaviour so far undertaken on feral animals, have been mainly on such island populations. Behavioural studies on herds of feral cattle and horses from various continents cannot be evaluated here because either no comparable observations have been carried out on the relevant wild forms in the natural state or such observations are impossible because the wild ancestors have died out. Feral dogs, cats, pigs, cattle and goats on the Galapagos Islands possess the full colour range of the domestic forms and reveal no changes in brain size. The Soay sheep of the St Kilda Islands, which returned to the wild very early on and have remained primitive, have the same small brain size as other domestic sheep. Cats that were released on the Kerguelens in the southern Indian Ocean about 1950 underwent no enlargement of the brain compared to other domestic cats in the few generations of the following 20 years. Domestic rabbits left there in 1874 multiplied so well that they eventually caused great damage to the vegetation. In contrast to the cats, they seem to have developed somewhat larger brains by 1900, though without reaching the level of the wild rabbit. To make a truly valid comparison, the brain size of the breed of origin would have to be known. Typical domestic animal colouration is found in the feral goats, sheep and pigs on the Hawaiian islands.

A striking exception to this persistence of typical domestic characteristics after several feral generations could be the case of the Porto Santo rabbits (Fig. 11.1). This form, from the island of Porto Santo near Madeira, is the smallest free-living rabbit, hardly any larger than a guinea

Fig. 11.1. Porto Santo rabbits, the smallest free-living type of rabbit, going back to animals at an unknown level of domestication from the early period of rabbit breeding.

pig, and may also be called the wildest of all rabbits living in the wild. Their particularly unruly behaviour makes it very difficult to purebreed them in captivity. To ensure the rearing of the young, domestic rabbits have to be used to suckle them. This is obviously a case of even lower psychogenic tolerance than that found in the wild animal, and is the opposite of the domestic animal. During the 500 years and more of their feral existence, the Porto Santo rabbit population underwent an evolution that led, so to speak, to a 'superwild' animal. The domestication level of a female with young from Portugal released in 1418 that founded this population is, however, unknown. They may quite well have been rabbits from runs at an early stage of transition from the wild to the domestic animal, in which the wild rabbit had been scarcely or only a little altered. So it is an open question whether one can speak of a really domesticated animal returned to the wild in this case.

How the success of returning to the wild depends on the level of domestication reached can be traced very well in the rabbit. From Roman times on, it was customary to keep rabbits in runs and this was continued for hunting purposes in the Middle Ages. At the same time, domestication probably progressed with rabbits kept in monasteries from the beginning of the Middle Ages and may well have concluded with the attainment of truly domesticated rabbits by the end of the Middle Ages. Even in the pre-Christian era in the region of Roman civilization, rabbits that had been kept in captivity were released on the Baleares Islands and Corsica. In the course of the Middle Ages, rabbits from runs were released in France, England and Germany, for example on the island of Amrun in the North Sea around 1230. All these rabbits led to modern populations of wild rabbits. So the original animals were wild ones kept in captivity that had not yet undergone any domestication.

In contrast to the rabbits spread by man during the Middle Ages, the vast majority of rabbits shipped and released around the world in modern times were obviously of the completely domesticated type. In 1859, 24 rabbits from England, presumably a mixture of wild and hutch-bred domestic ones, reached Australia and were released in the state of Victoria by a landholder. By 1890, Australia was inhabited by an estimated 20 million rabbits. Around 1930, Australia and New Zealand together exported about 100 million rabbit skins a year. The rabbits, which had become a plague in Australia, evidently remained domestic in their behaviour to a great extent, indicating that the wild animals in the colonizing founder population mostly underwent some genetic change by crossbreeding with the domestic part of the stock. They are described as digging less than genuine wild rabbits, even as climbing more, and the young are said to be found often in nests on and not below the ground. This would fit the domestic rabbit's behaviour according to the results of comparative behavioural observations of wild and domestic rabbits. The domestic rabbits released on the small North Sea island of Memmert near

Juist have also remained domestic in their behaviour (Fig. 4.3, p. 90).

The example of the rabbit underlines the inevitable conclusion that populations resulting from a return to the wild are more like wild animals the earlier in the course of domestication such a return to the wild takes place. In the initial stages of the development of domesticated forms from wild animals kept in runs in captivity, it would probably be difficult in general to determine whether the animals were still genuinely wild or already in the first stages of transition to the domestic animal.

Quantitative behavioural studies can help in the classification of living animals. Skulls, including those from archaeozoological finds, can be checked for brain size and possible alterations in the space for the sensory organs – always in relation only to the original ancestral geographically separate population. Even then classification is possible only on the presupposition that no selection has taken place on the characteristics in question in the meantime, and only in regions in which the wild species was originally absent; otherwise the problem becomes still more complicated. In such cases, interbreeding can only be recognized clearly when there is a characteristic that arose first during the process of domestication.

The moufflon populations of Sardinia and Corsica, from which the various moufflon populations inhabiting Europe today originated as a result of various releases, may serve as examples for such diagnostic problems. The lack of remains of this species from the time before human settlement of islands indicates that they are not indigenous wild sheep that came there over some Ice Age land connection with the Appenine peninsula. But if they were transported there by human agency in the course of the pre-Christian millennia, it might have been a matter of the release of completely wild animals or of animals in the very early stages of the transition to the domestic state. This latter interpretation seems quite feasible when the brain size of these moufflon is compared with the slightly higher values for moufflon from Asia Minor and the Near East.

Synopsis

Taming is not equivalent to domestication, nor is it a necessary condition for domestication. The contrast between free-living and tamed wild animals kept in captivity has its counterpart in the contrast between the feral and the tame domestic animal. Taming is, however, a precondition for confident and problem-free management of domestic animals. In male domestic animals, if this manageability is still not sufficiently attained by the selection that has taken place during the course of domestication, or by the additional process of habituation to humans (as is typical for taming), castration can lead to the desired result as a sort of further 'physiological taming'. The counterpart of taming in domestic animals is their returning to the wild. Populations living in the wild that have arisen in this way are as a rule characterized by a reduced environmental appreciation, higher psychogenic tolerance, and the typical behavioural organization of the domestic animal. Still, they behave more like wild animals the earlier in the initial stages of the transition to the domestic animal the return to the wild occurred.

12 | New domestications

In the preceding chapters, three principles of domestication were elaborated. (1) Wild animals seem to be more suitable for domestication the easier they are to breed successfully in captivity in crowded conditions (see Chapter 5). (2) Individuals from a species to be domesticated seem to be more suitable the smaller their relative brain size within that species (see Chapter 6). (3) The selection and combination of certain coat colour types can elicit direct domestication effects (see Chapter 8).

With the knowledge of these basic principles, it must now be possible to undertake purposeful selective and combinatory breeding for new domestications which enable the transition period between the wild and the domestic animal to be traversed within a few animal generations, in contrast to the domestication of animals of past millennia that proceeded slowly over hundreds of years and involved a great deal of chance. A further possible approach is to select rigorously for the typical domestic behavioural syndrome (the innate decline in environmental appreciation (cf. Chapters 4 and 5)), where the fundamental suitability for domestication seems to exist according to the first principle but no differences in populations or individuals corresponding to the other two principles can be defined.

An excellent example of this approach is provided by a study on breeding a fox that resembled the dog in its behaviour, undertaken by D.K. Belyaev and L.N. Trut. This experiment was carried out with silver foxes on a fur farm belonging to the Soviet Academy of Sciences near Novosibirsk. Young foxes aged between $1\frac{1}{2}$ and 2 months were selected according to the criteria of their tolerance of hand-feeding, their reaction to being handled by humans, and their response to being called. In this way, after 15 years of constant selection, foxes were finally bred that came when they were called, tolerated being petted and picked up by humans, wagged their tails in greeting and barked on seeing humans; in brief, they behaved in practically the same way as dogs. The corticosteroid level in the blood of these foxes was lower than that of the unselected animals in the original group. The foxes completely domesticated in this way were subjected to experimental stress produced by injections of adreno-corticotrophic hormone (ACTH, cf. Chapter 5) and by being spatially confined either by being put into a small carrying cage or by having their free movement hindered. In comparison to non-selected foxes treated in the same way, this led to a proportionally higher rise in the corticosteroids in the blood of selected animals to different degrees depending on the season, i.e. connected with alterations in sexual activity. The absolute corticosteroid level in the control animals was mostly still higher than in the 'domesticated' foxes. So in the course of domestication, the expected alteration in the functional state of the adrenal glands had occurred (cf. Chapter 5). In addition to their dog-like behaviour towards humans, the annual periodicity in reproduction was destabilized. The breeding season (oestrus) occurred earlier, and there was a tendency towards two seasons per year, as in the dog. So the influence of the geophysical time-synchronizer on sexuality had been reduced, as is typical for the domestic compared with the wild animal. Probably because of poor sexual co-ordination as a consequence of the temporal destabilization, there were fewer successful matings in the selection group, so that their reproductive success dropped. In addition there were disruptions in the mothers' milk production and in their behaviour towards their young, including cannibalism.

A similar case of significant reduction in reproductive success due to displacements in the breeding season, or rut, was observed in white red deer in a nature reserve in Bohemia. Their rut begins later in the year than that of normal-coloured deer and also lasts longer, so that the birth of the young sometimes takes place very late in the season: this reduces their chances of surviving the following winter. Practising selection for colour on these white deer, which go back to crossbreeding Bohemian stags with colour mutants from a Southwest Asian population around 1780, has quite unintentionally led to a first step towards a possible domestication of this species.

These observations of decreasing reproductive success due to de-stabilization of the breeding period imply a fundamental conclusion for the primary phase of domestication: as long as the attenuation of the seasonal effect of the time-synchronizer causes the temporal coordina-tion of oestrus of individuals in a small breeding group to be poorer than is normally the case in the wild animal, the stock of this primary founder group will increase only slowly. There seem to be two different ways of overcoming this initial problem. As the stock gradually becomes larger, the phases of some of the males and females will eventually converge by pure chance and so they will be able to reproduce without limitation. Moreover, further reduction in the influence of the seasonal time-synchronizer can involve the mating readiness of the males becoming more and more permanent, as is the case with many domestic animals, so that the difficulties in coordinating with the oestrus of the females disappear. The higher psychogenic tolerance of the domestic animals arising in this way and the consequent larger number of progeny through increase in the litter sizes and/or in the sequence of births should lead to a faster multiplication of the stock.

A new domestication of the field vole (*Microtus arvalis*), which can be considered as having already succeeded even though it was also probably unintentional, was carried out at the Agricultural College at Brno, Czechoslovakia, where an albino mutant was taken for breeding. After 30 to 40 generations, these albino animals were much heavier and larger than normal for their wild relatives, weighing between 50 and 60 g (Fig. 12.1). Like laboratory mice, they could be picked up without the protection of gloves and moved about in the cage. As described in the scheme given for the development in body size in the course of domestication in Chapter 5, these voles were especially small when breeding was begun and then increased in average size from generation to generation. The same thing happened with normally coloured field voles bred in captivity except that they did not become as large as the albinos. So, in this case, primary selection according to colour led to a favourable shift in the development of tolerance during the breeding process.

In the past few decades, domestication experiments have been undertaken on some large mammal species with the goal of obtaining animals for various uses. All these experiments (begun without knowledge of the basic principles worked out by the author) have in common that either they did not reach the phase of transition from the wild to the domestic animal but remained at the stage of tamed wild animals kept in captivity, or they only very slowly reached the first stage of a domestication.

In African colonies at the beginning of this century, an attempt was undertaken to domesticate the Burchell zebra (*Equus (Hippotigrus) quagga*) for use as a work animal. Except for some successes in taming, which have nothing to do with domestication, this failed. Now that this project is no longer of any interest there, it seems worth considering domesticating zebras for meat production in the Sudan and the sahel

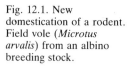

Fig. 12.1. New domestication of a rodent. Field vole (*Microtus arvalis*) from an albino breeding stock.

zone. The basic breeding material would be available, since the first stage of behavioural change in the direction of the domestic animal by the selection of 'white' animals has been demonstrated by the author's research group, and the foundation of a first breeding stock has been supported in the Georg von Opel Animal Park and Research Station in Kronberg. This has opened the way for entry into the transitional phase from the wild to the domestic animal (Fig. 12.2).

Elands (*Taurotragus oryx*) have been kept at Askania Nova in the South Ukrainian steppe since 1892 (Fig. 12.3). The whole stock, based on only five males and five females imported between 1892 and 1964, had produced 461 calves by 1967, although distinct inbreeding damage has appeared since about 1930. These large antelopes are kept like cattle in single stalls in stables in winter and in large enclosures or on open steppe pastures tended by mounted herdsmen in summer. The calves have been hand-reared since the end of World War II, so as to obtain tame animals that are manageable for milking, the main breeding purpose initially being to increase milk production. Eland milk has approximately three times the fat content and double the protein content of cow milk. The

Fig. 12.2. 'White' zebra foal with normal-coloured conspecifics; member of a breeding group of Burchell zebra at the beginning of the transition phase from the wild to the domestic state.

Fig. 12.3. Elands (*Taurofragus oryx*). Intensive endeavours were made to domesticate this animal for decades, mainly in Askania Nova in the Ukraine, but evidently no more was achieved than the initial level of domestication.

milk yield of the eland cows was different in the three genealogical lines in the herd and so is accessible to selection. Such selection was carried out on the basis of the cows' reactions to being milked, which is a decisive factor, since they do not all let themselves be milked. Up until 1971, the peak yield for an eland cow in one milking period was 637.9 kg, an average value of 2.5 kg daily, which is about a quarter to a fifth of the production of a cow from an improved breed of dairy cattle. So this domestication experiment has led to elands that can be milked and tended in herds, and subjected to further selection. The intensive taming of the calves by hand-rearing, not the attainment of tamer animals by a hereditary decline in environmental appreciation, is the factor of decisive importance. So at most only the initial phase of domestication has been entered. This basic phase, consisting of keeping the young in captivity and taming them, probably corresponds in principle to what occurred in Ancient Egypt, where other large antelopes, such as the oryx and addax, were kept without this resulting in genuine domestication. A comparison with the reindeer, which in some places has probably not progressed much beyond this stage of domestication, may also come to mind.

A further current domestication experiment also being carried out in the Soviet Union involves the elk or moose (*Alces alces*) from the deer family (Fig. 12.4), the only domestic animal from this family so far being the reindeer. Whereas the foundation for domestication is laid in the eland by its breeding well when kept in captive groups, it is precisely this factor which seems extremely problematic in the elk. It is not gregarious when living free and the size of the bands does not exceed four as a rule. It is hard to keep elk in captivity, habituation being especially difficult, and death due to stress is frequent in adults caught in the wild. Conspicuous

Fig. 12.4. Elk (*Alces alces*). So-called domestication experiments with this large northern deer have so far been confined mainly to intensive taming of young animals and so have nothing to do with domestication as such. Compared with some other deer species, the elk seems to be thoroughly unsuitable for domestication.

stress effects arise when the density of elk in the wild is too high. In the Oka Terrace Reserve, south of Moscow, the population rose from an average of five elk per 1000 ha to 68 per 100 ha between 1949 and 1959. In addition to enormous damage to the forest, this was accompanied by a decrease in the rate of progeny. Young male elk often had weak, irregular antlers and infestation with endoparasites increased. In other over-stocked regions, other signs of psychosocial stress became apparent. The life expectancy dropped, the animals reached sexual maturity late and non-pregnant females were encountered more frequently. This in turn leads to a decrease in stock, as has been found experimentally in rodent populations (cf. Chapter 5). This documents the very low psychogenic tolerance of the elk and makes it right from the start a most unsuitable candidate for genuine domestication. The experiments have been in progress in the Pechora-Ilych Nature Reserve since 1946. It proved to be most suitable to begin with newborn calves aged three days at the oldest. By hand-rearing these animals they could be intensively tamed and a relation similar to imprinting could be developed towards the person caring for them. This generated a reaction of following the keeper and paved the way for their later being allowed free movement in a pasture. They were accustomed to being massaged in the area of the hindlegs and belly early on so as to make them easier to milk later. Milking increased in production during a lactation period from the normal 150–200 l to a maximum of 429 l. Elk milk typically has a higher fat, protein and mineral salt content, but less lactose (milk sugar) than that of domestic cows. Using the elk to transport loads in the swampy forest areas of Siberia, one of the main goals of the domestication project, can also be achieved by training the young from the first months of their life. In this way, elks can be accustomed to carrying loads and drawing sleds. The maximum load an adult elk can carry is one-quarter of its body weight, which normally means between 80 and 100 kg – 130 kg at the most. The elk travels at an average speed of 4 km per hour under such a load. Fully grown elks from their fourth year onwards can draw a sled with a load of 300 to 400 kg for between 30 and 40 km per day. It is questionable whether this is profitable. This domestication experiment is mostly only an intensive taming of young animals, which does not really have anything to do with domestication as explained in the previous chapter. A first breakthrough to domestication could be achieved only when the psychogenically tolerant animals survive longer and also reproduce successfully at the husbandry station. This could lead to a very gradual establishment of innate decline in environmental appreciation but would have to be accompanied by producing the basic state required for domestication, the capability of being bred easily in captive groups, such as already exists in other species.

There can hardly be any claim of real success in genuine domestication so far in the two cases of the eland and the elk. Intensive taming begun early makes it possible to keep and use them but this does not directly involve domestication. Environmental appreciation by individual animals may have been impoverished by their being kept in captivity but it is not innately reduced as it is for domesticated animals. Even intensive taming of young animals for many generations does not of itself involve any selection towards domesticity; but it can provide a point of departure for this selection.

Recently, further approaches to keeping wild animals for use, partly coupled with ideas of domestication, have been undertaken with several species of large mammals. In South America, it was proposed that the giant rodent capybara (*Hydrochoerus hydrochaeris*) (Fig. 12.5), indigenous to the swamps, lakes and rivers there, might be used for meat production. The same purpose was envisaged for Barbary sheep (*Ammotragus lervia*) (Fig. 12.6) from the dry lands at the northern and southern fringes of the Sahara. An examination of the suitability of this

Fig. 12.5. The capybara (*Hydrochoerus hydrochaeris*), a giant South American rodent weighing up to 50 kg, seems to be a promising candidate for serious attempts at domestication. It is adapted to life in grasslands as well as in lakes, rivers, and swamps, and feeds on grass and aquatic plants, which makes a possible product of domestication particularly interesting for improving the protein supply in tropical regions.

Fig. 12.6. The Barbary sheep (*Ammotragus lervia*), a relative of the sheep and goats, has already been considered for meat production in dry areas in North Africa. Due to its high rate of reproduction even when kept in very confined conditions, its domestication seems both feasible and useful.

latter species for being kept on game farms and eventually domesticated was carried out by Claudia Visosky in the author's research group, and it seems it may have some potential. In recent years, species of the deer family have become important for venison production but also – particularly in the Soviet Union, China and lately also in New Zealand and Canada – for the velvet from their antlers for medicinal purposes. In Central Europe, mainly fallow deer (*Dama dama*) are kept wild on game farms, and in northern Europe both these and red deer (*Cervus elaphus*) are game-farmed (Figs. 12.7 and 12.8). Both these species are kept in New Zealand, the rusa or Timor deer (*C. timorensis*) on Mauritius, and sika deer (*C. nippon*) and red deer in the Soviet Union and China.

Red deer farming (sometimes called domestication in Scotland) is

Fig. 12.7. Père David's deer (*Elaphurus davidianus*) has turned up among the various species of deer that have been used in game farming for meat production only in a preliminary experimental stage in Scotland. This large Chinese deer is adapted to swampy ground and has been extinct in the wild for several centuries. It has only survived thanks to intensive efforts to breed it in captivity where it has multiplied well and could perhaps be used with a view to domestication in agriculture in certain regions.

Fig. 12.8. Wild red deer (*Cervus elaphus*) are kept in parks and enclosures in many countries for the production of venison and of velvet for medical purposes, but almost no serious attempts have been made at domestication, disregarding unintentional and unwitting entry into the transitional phase from the wild to the domestic state due to selective breeding for colour. As with the elk, keeping tamed animals in a halter as for a domestic animal has nothing to do with domestication.

based on hand-rearing or other taming of wild animals. The red deer readily learns to tolerate close human contact, which may even hinder effective selection during domestication. What is being achieved merely parallels the second type of selection in the very slowly proceeding early prehistoric domestication described in Chapter 5.

Keeping fallow deer in Central Europe has increased almost explosively. It is a way of exploiting grassland and fallow areas, no longer used for other traditional agricultural purposes, to gain additional income with little effort. The unused areas came about mainly as a result of the availability of well-paid employment in non-agricultural industries for people living on the land and the declining profitability of unfavourably situated or unfavourably structured areas (marginal yield areas). The search for an alternative to the classical domestic animals cattle and sheep for grazing wastelands, which was mainly carried out by Günter Reinken, showed that fallow deer were particularly favourable. The criteria for choice were long life, resistance to disease, winter hardiness, high fertility, ease of giving birth, good nature, low feed requirements, good feed utilization, early maturation, good meat growth, high meat yield and excellent meat quality. The demand for venison is high; 26 000 tonnes valued at just about DM139 million were imported into the Federal Republic of Germany in 1975/76 in addition to approximately 11 700 tonnes from animals killed in the country.

The slaughter yield of fallow deer is very good. It is defined as the percentage of the animal body's live weight remaining after subtraction of the slaughter loss such as head, feet, coat, intestines and blood, and amounts to an average of 57% for a one-year-old male deer. This matches the value of meat cattle and calves, while fattened lambs produce a slaughter yield of about 50%. Only pigs, with 70% to 75%, supply more. The weight proportion of the valuable cuts rich in meat, such as fillet, neck, back and haunch, is 57% on the average for fallow deer, about the same for fattened calves at between 55 and 60%, but much less (at between 45 and 50%) for fattened lambs and pigs. From this total, the percentage of the haunch cut from fallow deer is about 38% and so almost 19% higher than from lambs and pigs and in the same range as for calves. The annual meat production per hectare of pasture can be higher for fallow deer than for sheep and cattle. Since the price of venison is higher, the gross profit per hectare exceeds that for cattle and sheep.

Pasturing fallow deer is quite simple: a fence 1.80 m high is usually sufficient to keep them in. When grazing, fallow deer, in contrast to some other species, leave few rank patches, since they graze off almost the whole vegetation more or less uniformly. Fallow deer can also become very tame so that they are easy to attract from one part of an enclosure to another with feed. Nevertheless, they usually remain very easily alarmed so that they panic readily, which can lead to losses when they hit fences in attempting to flee. Domestication seems to be the only way of reducing this behavioural trait, which has negative consequences for keeping them in enclosures and catching them for slaughter and, because of acute stress, can lower the meat quality. This would involve a decline in environmental appreciation and reduction of stress that would not only result in domestic fallow deer being less easily alarmed than wild ones but would also improve their utilization of feed and their weight development, the certainty of sexual maturation in the second year of life, their reproductive success, and their resistance to infection. The social tolerance and so the number of animals that could be kept per hectare would be increased. The desired attenuation of reactivity by a decline in environmental appreciation should be reflected in lower investment costs for fencing and in increased productivity due to higher weights and increased reproductive performance. Moreover, domestication could eliminate various legal problems involved in keeping wild animals in enclosures, at least in the Federal Republic of Germany. However, fallow deer should only be domesticated so far that the transition phase from the wild to the domestic animal is just crossed, so that a primitive domestic deer breed is produced that still possesses sufficient physiological endurance (cf. Chapter 10) to ensure that it can be kept for extensive utilization of unused land with a minimum of effort. The development of improved breeds would run counter to this need and so again reduce the competitive worth of fallow deer and perhaps also distinctly impair the taste of the meat.

As regards the basic criterion of suitability for domestication, amenability to being kept and bred in captivity, fallow deer are certainly appropriate. As a rule, no problems arise in breeding them in enclosures. Fallow deer are gregarious and seem to be socially fairly compatible, as has been shown by Cornelia Rammelsberg in the author's research group. This requirement for domestication is not fulfilled by all species of the deer family. Just as an attempt to domesticate the elk (cf. above) seems extremely problematic, so the roe (*Capreolus capreolus*) can be excluded from the species that would at first sight seem to be suitable candidates. Other examples of species that seem to be unfavourable are the North American white-tailed deer (*Odocoileus virginianus*) and the sambar (*Cervus unicolor*). Not every species which is kept in enclosures for meat production is basically suitable for domestication. In this respect, the choice of fallow deer proves to be particularly favourable, being near the reindeer but above the red deer and the sika deer in terms of suitabiity.

The principle of selecting a species for domestication according to brain size seems inappropriate in the case of fallow deer. Although the data available for estimating the brain size of Mesopotamian fallow deer, the only other geographically separate population of this species besides the ones from Europe and Asia Minor, are still much too sparse, they do not yet indicate any conspicuous differences between the two populations in this respect. The origins of the entire modern population of European fallow deer can be traced to Asia Minor, since the species clearly did not survive the last Ice Age in Europe. Fallow deer were widespread as game in the Mediterranean region, even in the pre-Christian era, and remains have been found in the archaeozoological material from a site on Mallorca. The Romans also brought fallow deer into Central Europe, where the animals became popular in parks and enclosures over the centuries. That they have been kept for centuries in game parks has already opened the way to the transitional phase from the wild to the domestic animal. Where selection did not occur in a natural way, for example through the predation by the wolf and the lynx, but was carried out by gamekeepers in the service of resident rulers, types deviating from the wild colour were preserved and their numbers increased by positive selection. The first stage of the transitional phase from the wild to the domestic animal was even unknowingly entered by selective breeding of pure herds of a conspicuous colour in some parks, for instance the black herd in the Favoritepark near Ludwigsburg.

In the normal wild colouring, fallow deer have a summer coat of rusty brown with lighter, cream spots on their upper flanks and backs, and a cream stripe on their flanks. They have a dark dorsal stripe that is blackish towards the rear, the belly is creamy white, and the rear of the upper thigh, i.e. the buttock region, is white with a sickle-shaped border of black on either side in front of the rust brown of the upper leg. The short tail is white underneath and black on top in continuation of the dorsal stripe, contrasting conspicuously with the white buttocks (Fig. 12.9). In winter, the upper side becomes a darker greyish-brown, the spots almost disappear, and the paler parts become light greyish-brown, the contrast between the buttocks and tail being retained (Fig. 12.12).

Black fallow deer have a summer coat with a black upper side on which

Fig. 12.9. Intensive attempts to domesticate the fallow deer (*Dama dama*), a species that has attained special significance as livestock in agriculture in Europe, were directed by the author using the basic principles of domestication expounded in this book. The first step was the exclusion of the wild colour, characterized by somewhat indistinct pale-coloured spots on a rust-coloured coat and the black and white contrast between the buttocks and the tail.

the spots are still visible in places as greyish-brown. The underside is mouse-grey to dark greyish-brown but the contrast between buttocks and tail is almost completely lacking, since all the usually pale coloured parts are greyish-brown (Fig. 12.10). The winter coat of such animals is darker overall than that of the normal-coloured ones from whom they can be easily distinguished by the almost entire lack of contrast between buttocks and tail (Fig. 12.12). The pigment distribution gene with its *uniformity* allele leads to a general darkening and simultaneous blurring of all colour contrasts. As studies of pigment distribution in the single hairs and photometric measurement of the total amount of pigment show, this colour factor is comparable to the *agouti* locus with its *non-agouti* allele in other mammals (cf. Chapter 2). Like with the latter gene, the allele for colour uniformity of the fallow deer is recessive, i.e. it has an effect on the phenotype only when it is homozygous (doubled). So mating between black animals cannot produce any wild-coloured fawns, though mating between wild-coloured animals may very well result in black fawns.

White fallow deer do not completely lack pigment; they are not albinos. The fawns are beige with faint spots when born. During their first year this colour becomes paler, in winter it is a greyish-yellow and the creamy white colour of the adult animal is reached at the age of two years. The spots can then just be made out in oblique light, while the forehead is somewhat more strongly pigmented and so darker (Fig. 12.10). The winter coat of adult animals can darken to varying degrees to a light brown on the neck. The *white dilution* allele of the pigment locus responsible for this colour is also recessive; white animals have it doubled (homozygous). Fawns that later lighten to the white colour can be born

Fig. 12.10. Both black and white fallow deer display attenuated behaviour (thus pointing in the direction of domestic animals), which does not occur in wild-coloured types.

either from two white or from two wild-coloured parents, but mating between two white animals produces no wild-coloured progeny.

Another frequent colour type among fallow deer is the light colour called menil. Such animals have a summer coat that is light, glowing rust coloured on the upper side. The dorsal stripe is of the same basic colour. The spots are pure, glowing white as is the underside right up to the flank. The upper side of the tail and the borders of the rump are rusty brown (Fig. 12.11). The winter coat retains the contrast between the upper side and belly as well as the spots, which gives the impression of the summer coat of the normally coloured animals (Fig. 12.12). The hereditary characters involved affect the formation of the dark brown or black pigments, the eumelanins, and corresponds to the black pigment locus in other mammals, which alters black to brown by decreasing the eumelanins. Transitions to the wild colour and also to a much paler colouring are due to individual ageing and to wear at the hair tips towards autumn (Fig. 12.11), but also may suggest the participation of a second modifying gene.

A colour type that occurs only rarely if not specially bred combines a brown upper and lighter greyish-brown underside, with an even distribution of pigment, as in the black fallow deer. It is based on a combination of the *uniformity* allele of the pigment distribution locus with the eumalanin-reducing, i.e. *menil*, allele of the *black* locus.

Given these various coat colourings, the third principle of domestication

Fig. 12.11. In the so-called menil colouring of the fallow deer (left) the basic colour is light rust set off from the bright pure white spots and underside. The stripe down the middle of the back is not black, but rusty brown, as are the upper side of the tail and the borders of the buttocks. Even lighter coloured menil types (right) are found; these seem to result from genetic changes not yet well understood or from modifications during individual growth. Transitional stages from menil to the wild colouring are even more frequent. Compared with the behaviour of wild-coloured animals, the menil ones, like the black ones, seem to be less timid.

seemed to be applicable for domesticating the fallow deer. Observations of the behaviour of white and black individuals in comparison to those with wild colouration was carried out in the author's research group initially by Rainer Schaad. Black individuals, when compared directly with wild-coloured ones in groups of mixed colours, proved to be somewhat less shy, so they seemed tamer. From occasional observations in large hunting reserves, it seems that black animals tolerate a closer approach than do wild-coloured ones of comparable age. In a display enclosure where visitors can move freely among the animals and feed them, they come back up to the visitors sooner in the early spring before large crowds have begun to come. When taking food from the hand, they are less excited, as is shown by less raising of tails, and they behave more trustingly on the whole than do wild-coloured deer. It has been demonstrated statistically, that when feeding in a mixed group with normal-coloured animals, black forms looked up less frequently than the others. At the fences of smaller zoo enclosures, black deer react with less alarm to the public than do wild-coloured ones and keep a shorter flight distance in the enclosure. So the overall behaviour of black fallow deer with respect to readiness to take flight and alarm reactions seems to be attenuated. White animals are similar and, in addition, their social bonding is looser without there seeming to be any excluding reaction on the part of the group. This can readily be seen in wild-coloured or mixed wild-coloured and black groups with white individuals. In purely white

Fig. 12.12. In winter, the darker-shaded wild-coloured and black fallow deer can be most readily distinguished by the black and white contrast between the buttocks and the tail in the wild-coloured animals. The menil animals keep their lighter colour and bright belly in the winter coat so that they are easy to distinguish from the wild-coloured ones. In this large herd of fallow deer in a wildlife park, which is used to taking food out of the hands of the numerous visitors in summer, the tameness decreases strongly in winter, when there are few visitors, and gradually reappears in early spring. In the transitional period, the varying degrees of timidity in the different colour types have a certain screening effect in feeding experiments. As this photo shows, all the black (left) and menil (right) animals in the neighbourhood crowded around the feeder while the wild-coloured ones at first stayed further off.

groups the social bonding amongst fallow deer differs very widely according to the season and the food supply, stronger in winter and looser in summer. Behavioural synchronization is very close amongst fallow deer in small groups. When one animal gets up after a phase of lying down, the rest also get up within a few minutes, and, when one lies down, so do the rest. In groups consisting only of white animals, individuals might remain lying down for quite a long time while others grazed alongside. The social, transmissable effect of getting up and lying down seems to be distinctly lower. This confirms the interpretation that social bonding is looser in white fallow deer. There are still no lengthy, detailed behavioural observations comparing menil animals with wild-coloured ones. According to occasional observations, the former are classifiable as less timid than wild-coloured ones.

The way to domestication has been opened up by the selection of certain coat colourings. As was the case long ago with the building up of black or white deer herds in various parks for aesthetic reasons or out of interest in the unusual and conspicuous, this direction was initially taken quite unknowingly for menil animals at an experimental station directed by Günther Reinken. The only decisive consideration was that light-coloured coats would be easier to market than normal-coloured ones; at the same time this colour may clearly distinguish fallow deer in stocks kept for meat production from those living wild or in hunting reserves.

Because of this clear distinction, several fallow-deer keepers had already mentioned this choice. Such purely practical reasons may, of course, have played an equally significant role in colour selection in the course of Neolithic and later domestications, i.e. in the origin of our classical domestic animals, as did the selection of the unusual discussed in Chapter 9. Selective measures apart from this colour choice, which was really co-incidental from the viewpoint of the real goal, that have been introduced into fallow deer breeding have so far been concerned with selection for size, feed utilization, slaughter yield, birth of twins and non-seasonal rut. These are all traits that can be developed only over a long period of breeding but they should be attainable to a large extent within the framework selecting for a decline in environmental appreciation, i.e. for the domestic animal behavioural syndrome via a lowered stress level.

Chapter 8 concluded that combinations of alleles of individual colour factors deviating from the corresponding wild type can intensify or at least alter behaviour. So for domesticating fallow deer, an accelerating effect is to be expected if colour combinations based on colour alleles already involved in behavioural attenuation, are crossbred. Further-reaching, direct selection for behaviour should speed up the process even more. So the current status of knowledge on domestication can be directly translated into a strategy for new domestications.

A breeding program was set up on this basis to eliminate the problems involved in keeping fallow deer for meat production on unused land; it has been in progress since 1979 under the scientific direction of the author and in close co-operation with Ernst-Adolf Gaede in the Rhineland Palatinate Teaching and Experimental Institute of Livestock Husbandry at Neumühle. Contrary to what has hitherto been done in the so-called Scottish red deer domestication program, the goal of a primitive domestic deer has been attained here within only two fallow deer generations by applying the strictest behavioural outsider selection to the rigorously pre-selected founder animals and their offspring, and combining at the same time the pleiotropic colour alleles for black and menil to obtain a brown type in the new stock. This new domestic fallow deer combines being much less easily alarmed with good handling properties, high body weight and weight gain, and normal sexual maturity in the second year of life (Fig. 12.13). The success of this first modern large-scale domestication of a mammal has opened the way for further planned new domestications based on the overall concept presented here. Such projects may attain significance ranging from helping countries of the Third World to provide

Fig. 12.13. Primitive domestic fallow deer have been achieved after only two animal generations of strictest combined selection and breeding on the basis of initial colour and behaviour selection of the founder stock.

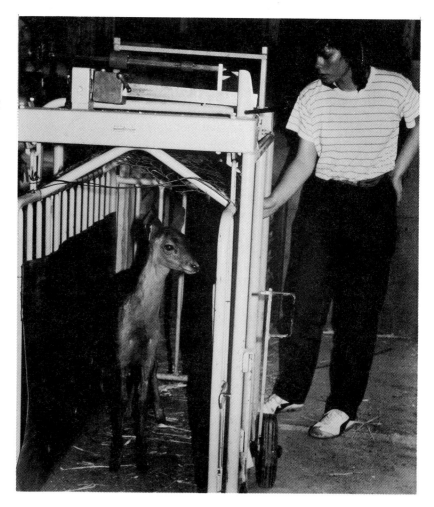

their own food supply to an influence on species protection and animal welfare in general by substituting new domestic animals for the wild species needed in medical research.

Synopsis

Attempts to domesticate large mammals undertaken in recent decades proceeded without knowledge of the basic principles of domestication explained here, so that these could not be applied practically. A model experiment on silver foxes with strict selection for behaviour over a period of 15 years demonstrated that the goal of a new domesticated animal, in this case a domesticated fox with behaviour comparable to that of the dog, can be achieved in principle.

Experiments with the eland and elk covering several decades and involving intensive taming of the young animals but without clear selection reached only the first stage of the way to the domestic animal. With red deer and fallow deer, the threshold to the transitional phase from the wild to the domestic animals has been crossed completely unintentionally in several cases independently of each other. Numerous attempts to keep and utilize several species of large mammals, mostly from the deer family, created the foundations for recent domestication experiments. In this context, a program using the basic principles of domestication, as outlined in this book, in a strategy to produce a domestic animal from the fallow deer has succeeded within two animal generations.

13 | Domestication and evolution

The evolution of organisms involves stochastic changes in the biochemical components that occur over long periods and are neutral with respect to selection, and equally stochastic shifts of allele frequencies in the dynamic development of populations. Most importantly, however, there is also continuous, selective adaptation of the relevant ability to compete or general ability to survive in changing environmental conditions. Human beings practising domestication intervene decisively in this natural process. A selection-neutral alteration to the biochemical systems can be neglected here in view of the time spans in question; stochastic shifts in allele frequencies play a predominant role when only the smallest founder population nuclei are taken over for breeding in captivity. Since only a very limited proportion of the whole gene pool, the totality of the genes of all individuals belonging to a population, is extracted and then multiplies a great deal again (genetic drift), the way the picture of the new domestic stock will differ from the ancestral wild population has a considerable element of chance. The decisive factor in domestication, however, is that the animals in question are prevented from choosing their own sexual partners and the effects of other mechanisms of natural selection are restricted (cf. Chapter 2). These are replaced by artifical selection, the determination of breeding according to human ideas and needs. In the last analysis, this is no more than a special form of natural evolution when the human species is viewed as a part of nature. When the transitional phase from the wild to the domestic state is entered, the changing environmental condition that usually elicits selective adaptation in the evolutionary process is the transfer from the natural environment, not human-influenced to any great extent, into the domain of people. This environment in the vicinity of humans enables creatures to occupy new, special ecological niches, which can be an important step for the survival and spread of populations. Two different types of special adaptation to humans can be distinguished: commensalism and domestication.

The decisive difference between these two is that in domestication the active partner subjecting the animal to his selective breeding is the human one, while in commensalism the animal actively penetrates the human environment and conquers it as a new habitat, frequently against human ideas and wishes. Commensalism literally means 'eating together with', that is that one creature feeds on the surplus food store or wastes of another, in this case of humans, without directly essentially damaging the host. Amongst the mammals, the house mouse, the brown rat, and the black rat are such commensals. The adaptation of a wild species to the close neighbourhood, the co-existence with humans, as a rule presupposes initial shifts in the complexes of timidity and curiosity, which are also affected by domestication. To build up and maintain populations successfully – even though humans more frequently appear in their living area and so increase the amount of stimulus load – it is usually necessary to raise the psychogenic tolerance, a situation resembling that of the domestic animal in principle. But in commensalism, tolerance is not necessary at any price.

The environmental appreciation of commensals has to be adaptively altered but it must not be subject to large-scale, general impoverishment, so that the animal does not become quite defenceless to being pursued by man, as are rats and mice in particular. This is precisely the case in feral domestic animals, which are easier to hunt than their wild relatives. Commensals must retain the efficiency of the information acquisition and processing system, the sensory organs and brain, to a large extent, since constant pursuit by humans requires the best possible perception and highest possible plasticity in behaviour. So differences occurring between the non-commensal and the commensal state probably chiefly concern the neurotransmitter system.

A connection with coat colour changes that would be conspicuous in their environment is unlikely, since the natural selection mechanisms are not neutralized or decisively reduced. Nevertheless, a primary colour shift does seem to be involved. It is conspicuous that in both house mice and black rats the colour on the belly of the commensals, or types capable of spreading considerably in commensalism, is darker, closer to the colour of the upper side – perhaps due to changes in the alleles of the *agouti* series – than the more or less white or pale-coloured bellies of non-commensal or only limitedly commensal forms of these species. Finally, in the commensal populations of the black rat spread all over the world, a genuinely black colour variant occurs regularly.

In contrast to commensalism, a decline in environmental appreciation – the increase in psychogenic tolerance occurring during domestication – can take place in any way since humans take over the protection and defence of the animals so that restricted capability to survive in the natural environment is of no consequence. The decisive factor in selection has become the human one. Breeding of the animal adapts it to a special ecological niche: the human environment. This enables a population that has passed through the transition from the wild to the domestic state to multiply more or less explosively under human support and control. This is accompanied by a spread into areas not previously colonized by the wild ancestors. Domestication raises the capability of survival to a completely new level and makes it possible easily to overcome ecological barriers to spreading. If the wild ancestors die out, due again to human activity or to unfavourable changes in their natural environment, as was the case with the aurochs or the wild horse in Europe and Soviet Central Asia, their domestic descendants preserve the evolutionary line in huge numbers. So the transition from the wild to the domestic state can become a factor of considerable, indeed decisive evolutionary significance.

This holds even for a preliminary stage in domestication, in which wild animals are merely caught, kept and transported. Being transported by human agency helps the animal to cross boundaries it had previously been unable to traverse on its own when it is released at the transport destination or escapes there. In this way, such species as fallow deer, sika deer, moufflon, raccoons, raccoon dogs, muskrats, and wild rabbits became part of the Central European fauna at various times. Wildlife management plays the same role: human intervention in the evolutionary process and the spread of animal species. In Central Europe, this is best exemplified by the red deer. For the purposes of so-called introduction of new blood and to produce larger antlers as more interesting hunting trophies, red deer, particularly from populations in the Caucasus and Altai but also from North America (Wapiti to Britain), have frequently been brought into European hunting-grounds and released there to interbreed with native stock. This has enriched the gene pool of the affected populations with foreign genes in a way that would never have occurred under natural conditions. However, this anthropogenic influence on breeding has nothing to do with selective breeding in domestication, since no alteration in environmental appreciation and so in behaviour was brought about.

Another intervention in natural selection is the additional feeding of red deer in winter practised in Central European countries. This greatly reduces the significance of climatic conditions as highly effective selection factors in cold, very snowy winters that normally reduce stocks. Increased survival of animals that would otherwise be naturally selected out raises the diversity in populations; less-well-adapted individuals get the chance to reproduce. Counterselection is then practised by hunters killing specimens clearly recognizable as deficiently developed. But, on the other hand, by shooting particularly strong bucks as trophies, hunters select against game of high survival capability. Natural selection pressure by predatory animals, namely the lynx and the wolf, has been eliminated by their extermination, which again contribute to higher propagation of weaker animals. So human influence has altered evolution in these populations in various ways so persistently that they can hardly still be regarded as natural, as long as one does not regard humans too as a 'natural' selection force. This form of exerting influence bears no relation to domestication. Wildlife management is a second way of radically altering the evolution of an animal species, alongside domestication.

If a species returns to the wild after passing the transitional phase from the wild to the domestic state but before reaching the stage of an improved breed, this may lead to the origin of a commensal from a primitive domestic animal. The less timid behaviour of such an animal towards humans and its reduced environmental appreciation enable it to live in high population densities in the direct neighbourhood of human settlements. Examples of this are feral primitive dogs especially in countries in the Near East and the tropics, which roam the outskirts of settlements and their rubbish dumps in search of edible refuse. Such dogs can be dangerous not only in mass occurrences but also above all as transmitters of rabies. By occurring in large herds, feral animals can also influence the evolution of other animals, due to their higher psychosocial tolerance and therefore destruction of the environment by overgrazing and trampling. This is particularly so for rabbits, sheep and goats, but it is also true for pigs. For instance, it became necessary to erect fences in Hawaiian national parks to keep out feral pigs, goats and sheep to protect the vegetation.

Normally the course of mammalian evolution is progressive, which is mainly attributable to the complexity of the brain and the sensory performance. The parallel, higher development of the brain is a phenomenon that can be observed independently in all groups of mammals. On the other hand, regressive evolution involving a reduction of a level of sensory capability, once reached without the decreasing performance of one sensory organ being compensated for by increasing performance in another and with a reduction of the central nervous system, is known from the development of sessile and parasitic forms in various lines of invertebrates. Since, however, the decline in environmental appreciation in the course of domestication involves just such a reduction, domestication may indeed be regarded as a special kind of regressive evolution. At least for the brain, it must be taken into account that domestication, where it began with species that were geographically widespread and in various stages of progression, was initiated with the least progressive populations in each case (cf. Chapters 3 and 6). The evolutionary level of the modern southern wolf, the primary ancestor of the dog, still corresponds to that of the wolves that inhabited Europe about a half a million years ago. Later progressive transformations occurring in the northern wolves did not take place in the relatively isolated southern populations. The specialized developments within the various geographically and ecologically separate populations of wild cats also began more than half a million years ago. In the populations from which the domestic cat originated, i.e. the wild cats of northeast Africa and Arabia as well as the wild cats of South Asia, no higher development of the brain took place, in contrast to all other populations of this cat group. The warty pigs, which, together with the also very primitive banded pigs from the group of Eurasian wild pigs, formed the basis for the domestic pig and now only exist in limited relict populations on the islands of Indonesia, are representatives of a basic evolutionary level of pig that was widespread in Eurasia at the end of the Pliocene and beginning of the Pleistocene and was only replaced by genuine wild pigs later than about one million years ago. The lack of progression in such primitive populations, in which hardly any change has occurred in relic occurrences for over a half a million to a million years, appears in each of these cases as a pre-adaptation to domestication. This makes it plausible that refuge areas of late Neogene and Lower Pleistocene faunal elements finally became the first domestication centres.

The Near East is one of these regions. The vicissitudes of climatic history in the Ice Age in the north of Europe and Asia led to animal species not especially well adapted to cold periodically becoming extinct in large areas. These areas were then re-colonized from residual populations, mainly those from the lands remaining warmer at the northern borders of the Mediterranean region and Central Asia. These populations were in turn driven out and advanced once more, whereby they were exposed to considerable selection pressure that accelerated their evolution. The environment in the Near East and South Asia, however, remained much more uniform during this whole period. Populations living there did not have to adapt so much to changed conditions as those further north and were in addition relatively isolated from their northern relatives by the geographical barriers of the Black Sea, the Caucasus Mountains, the Caspian Sea, and Soviet Central Asian deserts and the high mountains of Inner Asia. So the forms in the Near East were able to remain evolutionarily conservative while constant evolutionary change was taking place in the north. So, after the Ice Age, a large reservoir of comparatively unprogressive large mammals had collected in the Near Eastern region – a reservoir for important domestication.

On the basis of the food crisis concept in archaeology as propounded by Mark Nathan Cohen, the post-Pleistocene world-wide adoption of agricultural economies should be understood as the result of an ever-increasing human population pressure, reaching a critical point where strategies for increasing the food supply within the traditional hunting-gathering life-style had been exhausted. A new level of highly economic food acquisition, sustaining large sedentary communities had to be achieved where coincidentally, in addition to the ubiquitous plant cultivation, medium- to large-sized members of the regional fauna proved, a result of the first prehistoric attempts at game-farming, to have a clear disposition towards being domesticated. This situation arose in the Near East, where the populations of relevant species still represent archaic evolutionary grades.

Domestication as a special kind of regressive evolution also contains the seeds of speciation, i.e. the origin of new species. The formation of two or more new animal species from a common initial species presupposes in principle the development of isolating mechanisms which, with increasing efficiency, prevent mutual reproduction. Animal species are understood to be reproductive communities, the members of which can mate unrestrictedly with each other and produce unrestrictedly fertile progeny under natural conditions, i.e. not in captivity. Disturbances, mechanisms of mutual isolation, can limit both the free mating as well as the undisturbed development of the progeny and their fertility. Consequently pre-mating and post-mating isolation mechanisms are distinguished. In well-separated species of mammals, isolation factors from both these mechanisms or the pre-mating mechanism alone play a role. This is the case, for example, in the wolf, jackal and coyote species of the genus *Canis*, which can be crossbred with each other and produce fertile progeny at least via the wolf-descendant, the dog, although such matings do not occur at all under natural conditions, or only extremely rarely as in the case of the coyote and the dog. The North American bison (buffalo) and the European bison (wisent) are another pair of species that can be crossbred in captivity without restrictions.

Types of pre-mating isolation are: temporal isolation, which can involve both seasonal and diurnal shifts, ecological isolation in the form of different requirements with respect to environmental quality; social isolation based on behavioural differences; and structural isolation, which includes differences in bodily structure that make mating difficult. For domestic animals and their wild ancestors, it is initially to be expected that temporal isolation will exert a certain effect in principle. As explained as Chapter 4, temporal organization of behaviour is smoothed out in domestic animals. Their activity is distributed more uniformly over the day and, as far as reproduction is concerned, more uniformly over the seasons. The seasonal lengthening of the reproductive disposition, the weakening of its close connection to the seasons, in fact provides feral poulations in the distribution areas of their wild species with only periodic opportunities to mate with them. Mating is only possible in the limited reproductive periods of the wild species, despite the fact that the feral animals are in principle capable of unrestricted interbreeding with the wild form. The same holds good for the weakening of activity differences depending on the time of day in domestic animals. If the corresponding wild species has a strict activity peak, domestic animals ready to mate can, at least with increased probability, mate with each other during the rest of the day, but not with the wild animals that are either inactive or at least less active then. Demonstration of the isolating effect of such mechanisms has been provided by an experiment with a mixed population of wild mice and white laboratory mice. Judging from the main times of capture, the wild mice in this population had an activity peak in the evening and that of the laboratory mice was in the morning hours.

There are hardly any clues as to possible partial ecological isolation between domestic animals and their wild ancestors in the data available at present, if we rule out the tame domestic animal living directly in the human environment and the commensalism of feral dogs and cats (see above). Pariah dogs roaming at large in the Near East probably have habitats that overlap only partly with those of wolves; domestic cats roaming on the outskirts of settlements in Europe come into contact with wild cats only rarely.

The social isolation of domestic and wild animals is of fundamental importance. The differences in social behaviour described in Chapter 4, namely relaxation of social bonding and breakdown of social differentiation in domestic animals, partially connected with increased social tolerance, would lead one to expect direct inhibitory effects on the formation of mutual, mixed groups of domestic and wild animals. Dogs will scarcely be able to fit easily into wolf packs where complex and subtly differentiated social structures guarantee cohesion. Moufflon rams set free on Hawaii only joined groups of feral sheep when released singly, but if there were already several moufflon in the vicinity, small bands entirely of moufflon were formed so that separate social groups were created. Crossbreeds of moufflon and domestic sheep joined up with the domestic sheep. There is clear social isolation even in captivity between Porto Santo rabbits, which became feral at an early domestication level, and modern domestic rabbits. In a small, mixed population of wild and domestic rabbits synthetically created by E. Stodart and K. Myers, the domestic and wild animals, with the exception of a single wild male that joined the domestic group, separated into unmixed units. By contrast, female domestic rabbits released singly, but not males set free singly, were taken up into a wild population.

Structural isolation between wild and domestic animals is most to be expected where the size change during domestication has been so large that mating with the wild species is more difficult for this reason alone. This is mostly the case where individual domestic breeds have become particularly small compared to the wild ancestor, as for example in dogs, or particularly large, as for some breeds of horse. Difficulties can also occur when there are significant size differences between the various geographically separate populations of the wild species, as for instance in the wild sheep. Size differences of this kind would also involve mating difficulties between some breeds of domestic animals and some populations of wild animals, as exemplified by the Porto Santo rabbit that is too small to mate with wild rabbits from Central Europe.

Where feral animals and their wild relatives occur together, neither of the two types being present only as a single individual which could then only mate with partners of the other type (see the above-mentioned experiment with moufflon and domestic sheep), it can be expected that the mutual effect of all these isolation mechanisms will distinctly limit the probability of a purely chance mating between wild and domestic animals, even though there is no absolute barrier. An experimental demonstration of this is provided by the mixed population of wild and white laboratory mice assembled and studied by Rasbergen Reimov, Krystyna Adamczyk and Roman Andrejewski. Of a total of 18 albino females that were pregnant on re-capture, 16 had mated with albino males, as could easily be seen from the colour of their young. Two females produced both white and wild-coloured young so that there had evidently been double fertilization by albino and wild males. If one defines an isolation index as the ratio of hybrids actually found to the number of hybrids to be expected from purely chance mating, the result is a value in the region of 0.1. Such a value is at least twice as high as would be expected in well-separated species of mammals. On the other hand, it also proves the existence of thoroughly effective isolation mechanisms. This is matched in principle by the hybridization result in the small mixed population of wild and domestic rabbits mentioned above, where there were no hybrids in the 67 young born to wild females and only 15 amongst the 78 young born to domestic females. This puts the value of the isolation index at about 0.2. Findings from the Hawaiian population of moufflon and feral sheep as well as from populations of wild and feral goats on Aegean islands indicate similarly limited interbreeding. Correspondingly, without effective isolation mechanisms, European wild cats, after so many centuries of continuous opportunity for mating with stray domestic cats, would have displayed much more evidence of hybridization. Also, wolves in the Near East would have formed a more or less uniform mixed population with the parish dogs.

So domestic animals have started along a path of separate speciation with respect to their wild counterparts, even if this path has not led to complete separation. We are evidently confronted with an intermediate stage such as must always be traversed in the course of natural speciation or natural evolution. Intermediate stages of this kind always make it difficult to define whether they still belong to the original species or should be counted as a new, independent species and named accordingly in international zoological nomenclature. Introducing the concept of a 'semispecies' for such intermediate forms, which might alleviate conflicts, does help to differentiate but still necessitates an either/or decision when naming. Forms regarded as semispecies are usually given species names. If one treats domestic animals in the same way according to their evolutionary status, which is justifiable when seen from the viewpoint of the speciation initiated by domestication, then this possibility of assigning to them their own species names incidentally solves a much discussed problem of nomenclature at least in a practicable way.

Objections have often been raised to giving domestic animals the same names as their wild forms and to treating them in the nomenclature as if they were the equivalent of a subspecies, a geographically separate population of the wild species, especially when more than one wild species contributed to the origin of a particular domestic form. In this way, the earlier customary practice of treating domestic animals in the nomenclature as if they were separate species could be retained. So the domestic cat should be named *Felis catus* to distinguish it from the original wild cat *Felis silvestris*, the dog *Canis familiaris* in contrast to the wolf *Canis lupus*, the domestic horse *Equus caballus* as opposed to the wild horse *Equus ferus*, domestic cattle *Bos taurus* to distinguish them from the aurochs, *Bos primigenius*. By the same token, the names for other classical domestic animals (names for the wild animals in each case in Chapter 3) would be *Mustela furo* for the ferret, *Equus asinus* for the donkey, *Sus domesticus* for the pig, *Camelus bactrianus* for the domestic camel, *Camelus dromedarius* for the dromedary, *Lama glama* for the llama, *Lama pacos* for the alpaca, *Bos grunniens* for the domestic yak, *Bubalus bubalis* for the domestic buffalo, *Ovis aries* for the domestic sheep, *Capra hircus* for the domestic goat, and *Cavia porcellus* for the domestic guinea pig.

Synopsis

Evolution involves above all the continuous selective adaptation of survival capacity under changing environmental conditions. Humans supply the decisive selective factor in domestication, so it can be regarded either as standing apart from natural evolution or as a special case of it, depending on the point of view. The evolutionary transfer of a species from the natural into a purely human environment leads to commensalism when the species itself plays the active role and its environmental appreciation adapts without being impoverished to any large extent. It leads to domestication with decline in environmental appreciation at any cost when man is the active partner in the selective process. Domestication raises a species' survival capacity to a new level and allows it to overcome previous ecological barriers to its ease of spread with human aid. This evolutionary process may be regarded as a special case of regressive evolution, and begins with populations still at relatively low levels of evolutionary progression within their species.

Domestication is accompanied by incipient speciation characterized by the occurrence of pre-mating isolation mechanisms in the form of temporal and ecological isolation, social isolation and structural isolation. In particular, the mechanisms caused by the special behavioural syndrome of domestic animals restrict free interbreeding with the original wild species, even where feral animals occur alongside them.

14 | Overall synopsis

Domestic animals are kept and bred for constant use in and around the home. They satisfy human basic needs for food, clothing and warmth, supply raw materials for many other different products, help in performing work, are used in research for medical progress and occupy an important place in human social life. The course of history has been inseparably interwoven with the possession of domestic animals since the Neolithic, at least.

In contrast to wild animals, the appearance of domestic animals is very diverse especially with respect to the colour and type of coat, the body size and shape. This diversity is based on a reduction in natural selection on the one hand and selection in the course of breeding on the other and provides the foundation for the origin of numerous breeds. Primitive breeds are formed by crossbreeding with different populations of the wild ancestral species, via chance shifts in gene frequencies and through selective influences. Improved breeds, which are more easily defined, result from strict, standardizing breeding measures.

Becoming domesticated presupposes that a wild animal is suitable for the process and that people have an interest, for whatever reason, in keeping the animal. An endeavour to keep numerous species of animals in captivity, such as is demonstrable from the third millennium BC onwards at the latest, would not alone be sufficient to increase the stock of domestic animal species if the first prerequisite were not fulfilled. No clearly distinct period of increased domestications can be detected in the dates of origin of domestic animals, the spatial distribution of which is scattered widely over the millennia. However, these dates can only be determined with large ranges of uncertainty, in most cases due to the difficulties in distinguishing conclusively the wild from the primitive domestic animal remains in the archaeozoological data from the areas of primary domestication, areas moreover which have not yet been clarified for all species. The domestication dates for some species are still completely uncertain.

The behaviour of domestic animals seems to be weaker and less determined by environmental factors in contrast to that of wild animals. The environmental appreciation of the domestic animal is innately reduced when compared to that of wild animals. This is expressed in a lower intensity or even disappearance of particular patterns of behaviour, in a drop in motor activity, in more uniformity in the temporal organization of activity, in a loosening of social bonds and in a breakdown of social differentiation. A singular intensification of sexual activity contrasts with the general attenuation of other behaviour.

Under the same external conditions, genetic decline in environmental appreciation that determines the behaviour of domestic animals leads to increased psychogenic tolerance as compared with wild animals, i.e. to less stress. Stress is the state of general activation of an organism brought about by the influence of stimuli; it is a function of the stimuli impinging on the individual and the information gained from them, which is slight in the domestic animal. The seeming contradiction between a general attenuation of behaviour in domestic animals and the sole intensification of sexual activity is resolved by the stress concept. Since psychosocial stress resulting from living in a group plays a central role, wild animals that can be kept in groups in confined conditions are the most suited to domestication. Over the course of numerous generations, influences on growth and reproductive success, tending in different directions and arising from different keeping conditions and different selective measures applied by the keeper, lead to different stock developments typical of wild animals kept in zoos or of domestication.

The quality of the systems for information acquisition and processing responsible for psychogenic stress via the diversity of environmental appreciation changes in the expected direction towards a decline in environmental appreciation in the domestic animals during domestication. Reductions in the information acquisition system (the sensory organs), affect the eye, the ear and the olfactory organs to varying qualitative and quantitative degrees. In the information-processing system (the central nervous system), the neocortex of the cerebrum is the decisive area for memory capacity and complexity of interconnections. Since its development is connected with that of the whole brain, even measurements of the whole brain provide insight into the relative efficiency of this system, in addition to what can be learned from studies of special cortical areas. Comparing wild and domestic animals usually reveals decreases in brain size that may have been of a considerable extent during domestication. Where there are distinct differences in the brain sizes of a wild species, domestication began right from the start with the populations having the smallest brain sizes. So the smaller the brain size of individuals from a species to be domesticated, the more suitable they seem to be for this purpose.

The dynamic balance of the excitation transmitter substances in the nervous system, the neurotransmitters, is of decisive significance for what happens in information processing and so for environmental appreciation. Psychoactive drugs that influence the operational mechanisms of the neurotransmitters at the synapses alter this balance in various ways. This opens up the possibility of experimentally simulating, psychopharmacologically, the attenuation in information processing which has been postulated as the cause of observed alterations in behaviour occurring during the transition from the wild to the domestic state. An experiment of this type carried out with the cotton rat as a model of the wild animal confirmed the concept of the alteration of a complex consisting of information processing, stress and behaviour.

The coat colour of a mammal is related to the basic level of its activity, its reaction intensity and its environmental appreciation. The reason for this is probably to be found in the fact that up to a certain stage the pigments that determine colour – the melanins – and the catecholamine group of neurotransmitters that are to a large extent the basis of the information-processing system share a common biochemical synthesis pathway. Selection of certain coat colours can produce a behavioural change with a corresponding change in the stress system either towards attenuated behaviour and increased tolerance or in the opposite direction. Combinations of the alleles of single colour genes that deviate from the corresponding wild-type increase or alter their effect on behaviour. It follows that the strategy of selecting and combining certain coat colour types can produce direct effects on domestication.

The selection of coat colour variants distinctly deviating from the wild norm seems to be largely connected with human interest in the unusual. This readily assigns such particularly conspicuous animals a role in the cults of prehistoric and early civilizations. So the motive for domestication is often to be sought in cultural interests and the desire to acquire something unusual; no immediate, purely practical purpose in everyday life need have been involved. At the same time, the special amenability of many deviantly coloured animals for domestication set animal-keeping, which was begun independently of any secular considerations, on the road to success. In the same way, such a method may have accelerated the introduction of improved breeding of animals from the mass of still hardly altered primitive domestic animals. Selection for colour seems to be a central factor in domestication generally.

The decline in the environmental appreciation of domestic animals is accompanied by a lowering in their adaptability to acute stimulus loads due to functional alterations in the adrenal gland. This innate and breed-specific situation can be further intensified by special rearing methods in stimulus-impoverished environments. In this way, the spatial requirement for an individual animal may be lowered to close to the minimum set by the physical size in rationalized keeping of breeds improved for meat production. However, the limit of susceptibility, of succumbing to unfamiliar is reached at the same time. Special directions of selection for various utilization purposes have resulted in improved breeds of domestic animals whose capability of independent survival outside of the largely standardized environmental conditions provided for them is severely limited in many cases.

Taming is not equivalent to domestication, nor is it a necessary condition for domestication. The contrast between free-living and tamed wild animals kept in captivity has its counterpart in the contrast between the feral and the tame domestic animal. Taming is, however, a precondition for confident and problem-free management of domestic animals. In male domestic animals, if this manageability is still not sufficiently attained by the selection that has taken place during the course of domestication, or by the additional process of habituation to humans (as is typical for taming) castration can lead to the desired result as a sort of further 'physiological taming'. The counterpart of taming in domestic animals is their returning to the wild. Populations living in the wild that have arisen in this way are as a rule characterized by a reduced environmental appreciation, higher psychogenic tolerance, and the typical behavioural organization of the domestic animal. Still, they behave more like wild animals the earlier in the initial stages of the transition to the domestic animal the return to the wild occurred.

Attempts to domesticate large mammals undertaken in recent decades proceeded without knowledge of the basic principles of domestication explained here, so that these could not be applied practically. A model experiment on silver foxes with strict selection for behaviour over a period of 15 years demonstrated that the goal of a new domesticated animal, in this case a domesticated fox with behaviour comparable to that of the dog, can be achieved in principle.

Experiments with the eland and elk covering several decades and involving intensive taming of the young animals but without clear selection reached only the first stage of the way to the domestic animal. With red deer and fallow deer, the threshold to the transitional phase from the wild to the domestic animals has been crossed completely unintentionally in several cases independently of each other. Numerous attempts to keep and utilize several species of large mammals, mostly from the deer family, created the foundations for recent domestication experiments. In this context, a program using the basic principles of domestication, as outlined in this book, in a strategy to produce a domestic animal from the fallow deer has succeeded within two animal generations.

Evolution involves above all the continuous selective adaptation of survival capacity under changing environmental conditions. Humans supply the decisive selective factor in domestication, so it can be regarded either as standing apart from natural evolution or as a special case of it, depending on the point of view. The evolutionary transfer of a species from the natural into a purely human environment leads to commensalism when the species itself plays the active role and its environmental appreciation adapts without being impoverished to any large extent. It leads to domestication with a decline in environmental appreciation at any cost when man is the active partner in the selective process. Domestication raises a species' survival capacity to a new level and allows it to overcome previous ecological barriers to its ease of spread with human aid. This evolutionary process may be regarded as a special case of regressive evolution, and begins with populations still at relatively low levels of evolutionary progression within their species. Domestication is accompanied by incipient speciation characterized by the occurrence of pre-mating isolation mechanisms in the form of temporal and ecological isolation, social isolation and structural isolation. In particular, the mechanisms caused by the special behavioural syndrome of domestic animals restrict free interbreeding with the original wild species, even where feral animals occur alongside them.

Selected reading

General works, and references for Chapters 1 to 3

Alderson, L. (1978). *The Chance to Survive – Rare Breeds in a Changing World*. London: Cameron & Tayleur.

Antonius, O. (1922). *Grundzüge einer Stammesgeschichte der Haustiere*. Jena: Fischer.

Bökönyi, S. (1974). *History of Domestic Mammals in Central and Eastern Europe*. Budapest: Akadémiai Kiadó.

Brentjes, B. (1975). *Die Erfindung des Haustieres*. Leipzig, Jena, Berlin: Urania.

Brune, H. (1968). Rohstoffe der Haussäugetiere. In *Handbuch der Zoologie*, vol. VII, part 12 (4), pp. 1–47. Berlin: De Gruyter.

Clutton-Brock, J. (1981). *Domesticated Animals from Early Times*. London and Kingsville, TX: British Museum (Natural History) and Texas A & M University Press.

Epstein, H. (1971*a*). *Domestic Animals of China*. New York: African Publishing Corporation.

(1971*b*). *The Origin of the Domestic Animals in Africa*. New York, London, Munich: African Publishing Corporation.

Hagedoorn, A. L. (1954). *Animal Breeding*. London: Crosby Lockwood & Son.

Hammond, J., Johansson, I. & Haring, F. (ed.) (1958–1961). *Handbuch der Tierzüchtung*. Hamburg, Berlin: Parey.

Herre, W. (1955). *Das Ren als Haustier*. Leipzig: Geest & Portig.

Herre, W. & Röhrs, M. (1973). *Haustiere – zoologisch gesehen*. Stuttgart: Fischer.

Leeds, A. & Vayda, A. P. (ed.) (1965). *Man, Culture, and Animals*. Washington, DC: American Association for the Advancement of the Sciences.

Little, C. C. (1957). *The Inheritance of Coat Colour in Dogs*. Ithaca, NY: Cornell University Press.

Mason, I. L. (ed.) (1984). *Evolution of Domesticated Animals*. London: Longman.

Matolcsi, J. (ed.) (1973). *Domestikationsforschung und Geschichte der Haustiere*. Budapest: Akadémiai Kiadó.

Nachtsheim, H. & Stengel, H. (1977). *Vom Wildtier zum Haustier*. Berlin, Hamburg: Parey.

Searle, A. G. (1968). *Comparative Genetics of Coat Colour in Mammals*. London, New York: Logos Press/Academic Press.

Serpell, J. (1986). *In the Company of Animals*. Oxford: Blackwell.

Zeuner, F. E. (1967). *Geschichte der Haustiere*. Munich: Bayerischer Landwirtschaftsverlag.

Additional selected references for Chapter 3

Clutton-Brock, J. (1977). Man-made Dogs. *Science*, **197**, 1340–2.

Groves, C. P. (1981*a*). Systematic relationships in the Bovini (Artiodactyla, Bovidae). *Zeitschrift für zoologische Systematik und Evolutionsforschung*, **19**, 264–78.

    (1981*b*). Ancestors for the pigs: taxonomy and phylogeny of the genus *Sus*. *Technical Bulletin*, **3**, 1–96. Canberra: Department of Prehistory, Research School of Pacific Studies, Australian National University.

Groves, C. P., Ziccardi, F. & Toschi, A. (1966). Sull'asino selvatico africano. *Ricerche di Zoologia Applicata alla Caccia, Bologna*, suppl. **5**, 1–30.

Harrington, F. H. & Paquet, P. C. (eds.). *Wolves of the World*. Park Ridge, NJ: Noyes Publ.

Hemmer, H. (1975*a*). Zur Abstammung des Haushundes und zur Veränderung der relativen Hirngröße bei der Domestikation. *Zoologische Beiträge*, NF **21**, 97–104.

    (1975*b*). Zur Herkunft des Alpakas. *Zeitschrift des Kölner Zoo*, **18**, 59–66.

    (1975*c*). Zur Abstammung der Hauskatze (*Felis silvestris* f. *catus*): Sind Siamkatzen und Perserkatzen polyphyletischen Ursprungs? *Säugetierkundliche Mitteilungen*, **24**, 184–92.

    (1976). Zum Problem der Herkunft des Alpakas (*Lama* sp. f. *pacos*). *Säugetierkundliche Mitteilungen*, **24**, 193–200.

Jewell, P. A., Milner, C. & Boyd, J. M. (eds.) (1974). *Island Survivors – the Ecology of the Soay Sheep of St Kilda*. London: Athlone Press.

Meijer, W. C. P. (1962). *Das Balirind*. Wittenberg: Ziemsen.

Mengel, R. M. (1971). A study of dog–coyote hybrids and implications concerning hybridization in *Canis*. *Journal of Mammalogy*, **52**, 316–36.

Nadler, C., Korobitsina, K., Hoffmann, R. & Vorontsov, N. (1973). Cytogenic differentiation, geographic distribution, and domestication in Palearctic sheep (*Ovis*). *Zeitschrift für Säugetierkunde*, **38**, 109–25.

National Research Council (1983). *Little-known Asian Animals with a Promising Economic Future*. Washington, DC: National Academy Press.

Nobis, G. (1981). *Vom Wildpferd zum Hauspferd*. Fundamenta, Reihe B, Bnd 6. Köln, Wien: Böhlau.

Olsen, S. J. & Olsen, J. W. (1977). The Chinese wolf, ancestor of New World dogs. *Science*, **197**, 533–5.

Poplin, F. (1979). Origine du mouflon de Corse dans une nouvelle perspective paléontologique: par marronage. *Annales de génétique et de sélection animale*, **11**, 133–43.

Rempe, U. (1970). Morphometrische Untersuchungen an Iltisschädeln zur Klärung der Verwandtschaft von Steppeniltis, Waldiltis und Frettchen. *Zeitschrift für wissenschaftliche Zoologie*, **180**, 185–366.

Richter, C. P. (1954). The effects of domestication and selection on the behavior of the Norway rat. *Journal of the National Cancer Institute*, **15**, 727–38.

Röhrs, M. & Ebinger, P. (1980). Wolfsunterarten mit verschiedenen Cephalisationsstufen? *Zeitschrift für zoologische Systematik und Evolutionsforschung*, **18**, 152–6.

Röhrs, M. & Ebinger, P. (1983). Noch einmal: Wölfe mit unterschiedlichen

Cephalisationsstufen? *Zeitschrift für zoologische Systematik und Evolutionsforschung*, **21**, 314–18.

Stockhaus, K. (1965). Metrische Untersuchungen an Schädeln von Wölfen und Hunden. *Zeitschrift für Zoologische Systematik und Evolutionsforschung*, **3**, 157–258.

Tisdell, C. A. (1982). *Wild pigs: Environmental Pest or Economic Resource?* Sydney: Pergamon Press

Selected references for Chapter 4

Hart, B. L. (1985). *The Behaviour of Domestic Animals*. New York: Freeman.

Herre, W. (1979). Bemerkungen zur Evolution von 'Sprachen' bei Säugetieren. Zur Variabilität innerartlicher Kommunikation bei Caniden. *Zeitschrift für zoologische Systematik und Evolutionsforschung*, **17**, 151–73.

Leyhausen, P. (1962). Domestikationsbedingte Verhaltenseigentümlichkeiten der Hauskatze. *Zeitschrift für Tierzüchtung und Züchtungsbiologie*, **77**, 191–7.

Lorenz, K. (1959). Psychologie und Stammesgeschichte. In *Die Evolution der Organismen*, 2nd edn, vol. 1, ed. G. Heberer, pp. 131–72. Stuttgart: Fischer.

Pees, W. & Hemmer, H. (1980). Hirngröße und Aktivität bei Wildschafen und Hausschafen (Gattung *Ovis*). *Säugetierkundliche Mitteilungen*, **28**, 39–45.

Pilters, H. (1954). Untersuchungen über angeborene Verhaltensweisen bei Tylopoden, unter besonderer Berücksichtigung der neuweltlichen Formen. *Zeitschrift für Tierpsychologie*, **11**, 213–303.

Röhrs, M. & Kruska, D. (1969). Der Einfluß der Domestikation auf das Zentralnervensystem und Verhalten von Schweinen. *Deutsche tierärztliche Wissenschaft*, **76**, 514–18.

Scott, J. P. (1954). The effects of selection and domestication upon the behavior of the dog. *Journal of the National Cancer Institute*, **15**, 739–58.

Scott, J. P. & Fuller, J. L. (1965). *Genetics and the Social Behaviour of the Dog*. Chicago: University of Chicago Press.

Stahnke, A. (1987). Verhaltensunterschiede zwischen Wild- und Hausmeerschweinchen. *Zeitschrift für Säugetierkunde*, **52**, 294–307.

Stolte, H.-A. (1950). Über Entwicklung und Vererbung des Temperamentes wilder und domestizierter Kaninchen. In *Neue Ergebnisse und Probleme der Zoologie*, ed. W. Herre, Zoologischer Anzeiger, Erg. Bd. zu 145, pp. 980–99. Leipzig: Geest & Portig.

Woodward, S. L. (1979). The social system of feral asses (*Equus asinus*). *Zeitschrift für Tierpsychologie*, **49**, 304–16.

Selected references for Chapter 5

Axelrod, J. & Reisine, T. D. (1984). Stress hormones: their interaction and regulation. *Science*, **224**, 452–9.

Crandall, L. S. (1964). *Management of Wild Mammals in Captivity*. Chicago, London: University of Chicago Press.

Christian, J. J. & Davis, D. E. (1964). Endocrines, behavior and population. *Science*, **146**, 1550–60.

Ebinger, P. (1972). Vergleichend-quantitative Untersuchungen an Wild- und Laborratten. *Zeitschrift für Tierzüchtung und Züchtungsbiologie*, **89**, 34–57.

Gorgas, K. (1967). Vergleichende Studien zur Morphologie, mikroskopischen Anatomie und Histochemie der Nebennieren von Chinchilloidea and Cavioidea (Caviomorpha Wood 1955). *Zeitschrift für wissenschaftliche Zoologie*, **175**, 54–236.

Holst, D. v. (1969). Sozialer Streß bei Tupajas (*Tupaia belangeri*). *Zeitschrift für vergleichende Physiologie*, **63**, 1–58.

Jeche, M. (1980). Zur Bedeutung soziologischer Faktoren in der Tierhaltung und der Rolle der Nebennierenrindenfunktion. *Der zoologische Garten*, N F **50**, 337–44.

Lee, A. K., Bradley, A. J. & Braithwaite, R. W. (1977). Corticosteroid levels and male mortality in *Antechinus stuartii*. In *The Biology of Marsupials*, ed. B. Stonehouse & D. Gilmore, pp. 209–20. London: Macmillan Press.

Sassenrath, E. N. (1979). Increased adrenal responsiveness related to social stress in rhesus monkeys. *Hormones and Behavior*, **1**, 283–98.

Selected references for Chapter 6

Ebinger, P. 1974). A cytoarchitectonic volumetric comparison of brains in wild and domestic sheep. *Zeitschrift für Anatomie und Entwicklungsgeschichte*, **144**, 267–302.

Elias, H. & Schwartz, D. (1971). Cerebro-cortical surface areas, volumes, lengths of gyri and their interdependence in mammals, including man. *Zeitschrift für Säugetierkunde*, **36**, 147–63.

Frank, H. (1980). Evolution of canine information processing under conditions of natural and artificial selection. *Zeitschrift für Tierpsychologie*, **53**, 389–99.

Frick, H. & Nord, H. J. (1963). Domestikation und Hirngewicht. *Anatomischer Anzeiger*, **113**, 307–16.

Gorgas, M. (1966). Betrachtung zur Hirnschädelkapazität zentralasiatischer Wildsäugetiere und ihrer Hausformen. *Zoologischer Anzeiger*, **176**, 227–35.

Hemmer, H. (1972). Hirngrößenvariation im *Felis silvestris*-Kreis. *Experientia*, **28**, 271–2.

(1978*a*). Geographische Variation der Hirngröße im *Sus scrofa*- und *Sus verrucosus*-Kreis (Beitrag zum Problem der Schweinedomestikation). *Spixiana*, **1**, 309–20.

(1978*b*). Innerartliche Unterschiede der relativen Hirngröße und ihr Wandel vom Wildtier zum Haustier. Ein Diskussionsbeitrag. *Säugetierkundliche Mitteilungen*, **26**, 312–17.

Herre, W. & Thiede, U. (1965). Studien an Gehirnen südamerikanischer Tylopoden. *Zoologisches Jahrbuch, Anatomie*, **81**, 155–76.

Holloway, R. L. (1968). The evolution of the primate brain: some aspects of quantitative relations. *Brain Research*, **7**, 121–72.

Kruska, D. (1970). Vergleichend cytoarchitektonische Untersuchungen an Gehirnen von Wild- und Hausschweinen. *Zeitschrift für Anatomie und Entwicklungsgeschichte*, **131**, 291–324.

(1973). Cerebralisation, Hirnevolution und domestikationsbedingte Hirngrößenänderungen innerhalb der Ordnung Perissodactyla Owen, 1848 und ein Vergleich mit der Ordnung Artiodactyla Owen, 1848. *Zeitschrift für zoologische Systematik und Evolutionsforschung*, **11**, 81–103.

(1980). Domestikationsbedingte Hirngrößenänderungen bei Säugetieren. *Zeitschrift für zoologische Systematik und Evolutionsforschung*, **18**, 161–95.

Kruska, D. & Stephan, H. (1973). Volumenvergleich allokortikaler Hirnzentren bei Wild- und Hausschweinen. *Acta anatomica*, **84**, 387–415.

Klatt, B. (1912). Über die Veränderung der Schädelkapazität in der Domestikation. *Sitzungsberichte der Gesellschaft naturforschender Freunde zu Berlin*, **3**, 153–79.

Lüps, P. (1974). Biometrische Untersuchungen an der Schädelbasis des Haushundes. *Zoologischer Anzeiger*, **192**, 383–413.

Lüps, P. & Huber, W. (1971). Haushunde mit geringer Hirnschädelkapazität. *Mitteilungen der Naturforschender Gesellschaft Bern*, NF **28**, 16–22.

Moeller, H. (1975). Zur Kenntnis der Größenabhängigkeit von Hirnmerkmalen bei Hauskaninchen (*Oryctolagus cuniculus* forma domestica). *Zoologisches Jahrbuch, Anatomie*, **94**, 161–99.

Pirlot, P. (1974). Size and activity of brain-structures. *Zeitschrift für zoologische Systematik und Evolutionsforschung*, **12**, 152–5.

Weidemann, W. (1970). Die Beziehung von Hirngewicht und Körpergewicht bei Wölfen und Pudeln sowie deren Kreuzungsgenerationen N_1 und N_2. *Zeitschrift für Säugetierkunde*, **35**, 238–47.

Selected references for Chapter 7

Hemmer, H. (1970). Das Verhalten zu Artgenossen im Verlauf der Individualentwicklung der Baumwollratte (*Sigmodon hispidus* Say et Ord, 1825). *Zoologischer Anzeiger*, Supplementband **33**, *Verhandlungen der Zoologischen Gesellschaft 1969*, 306–11.

(1976). Man's strategy in domestication – a synthesis of new research trends. *Experientia*, **32**, 663–6.

Kreiskott, H. (1979). *Erregungszustände von Tier und Mensch*. Stuttgart, New York: Fischer.

Ng, L. K. Y., Marsden, H. M., Colburn, R. W. & Thoa, W. B. (1973) Population density and social pathology in mice. Differences in catecholamine metabolism associated with differences in behaviour. *Brain Research*, **59**, 323–30.

Raab, A. & Storz, H. (1976). A long term study on the impact of sociopsychic stress in tree-shrews (*Tupaia belangeri*) on central and peripheral tyrosine hydroxylase activity. *Journal of Comparative Physiology*, **108**, 115–31.

Schade, J. P. (1969). *Die Funktion des Nervensystems*. Stuttgart: Fischer.

Stevens, J., Livermore, A. & Cronan, J. (1977). Effects of deafening and blindfolding on amphetamine induced stereotypy in the cat. *Physiology and Behaviour*, **18**, 809–12.

Selected references for Chapter 8

Bernhard, W. (1965). Psychische Korrelate der Augen- und Haarfarbe und ihre Bedeutung für die Sozialanthropologie. *Homo*, **16**, 1–31.

Eibl-Eibesfeldt, I. (1950). Beiträge zur Biologie der Haus- und Ährenmaus nebst einigen Beobachtungen an anderen Nagern. *Zeitschrift für Tierpsychologie*, **7**, 558–87.

Fuller, J. L. (1967). Effects of the *albino* gene upon behaviour of mice. *Animal Behaviour*, **15**, 467–70.

Hemmer, H. (1978). Zusammenhänge zwischen Fellfarbe und Ausprägung des Aktivitätsrhythmus bei Labormäusen und wilden Hausmäusen. *Säugetierkundliche Mitteilungen*, **26**, 256–9.

Keeler, C. E. (1942). The association of the *black* (*non-agouti*) gene with behaviour. *Journal of Heredity*, **33**, 371–84.

Keeler, C. E. (1947). Coat colour, physique and temperament. *Journal of Heredity*, **38**, 271–7.

Keeler, C. (1961). The detection and interaction of body size factors among ranch-bred mink. *Bulletin of the Georgia Academy of Sciences*, **19**, 22–60.

(1975). Genetics of behavior variations in color phases of the red fox. In *The Wild Canids*, ed. M. W. Fox, pp. 399–413. New York, Cincinnati, Toronto, London, Melbourne: Van Nostrand Reinhold.

Keeler, C., Asteniza, J. & Fromm, E. (1964). Psychosomatics of fear in foxes. *Bulletin of the Georgia Academy of Sciences*, **22**, 64–71.

Keeler, C. E. & King, H. D. (1942). Multiple effects of coat color genes in the Norway rat, with special references to temperament and domestication. *Journal of Comparative Psychology*, **34**, 241–50.

Keeler, C. & Moore, L. (1960). Correlations between coat color, body size and behavior in ranch bred mink. *Bulletin of the Georgia Academy of Sciences*, **18**, 30–5.

Keeler, C. & Moore, L. (1960). Pigment gene effects on bodily proportions in mink. *Bulletin of the Georgia Academy of Sciences*, **18**, 69–84.

Keeler, C. & Moore, L. (1961). Psychosomatic synthesis of behavior trends in the taming of mink. *Bulletin of the Georgia Academy of Sciences*, **19**, 66–74.

Schwabe, H. W. (1979). Öko-ethologische Studien zur Ausbreitungspotenz der Hausratte (*Rattus rattus* L.). *Zoologisches Jahrbuch Systematik*, **106**, 124–68.

(1979). Pigmentationskorrelierte Verhaltensunterschiede bei Hausratten (*Rattus rattus* L.). *Zoologisches Jahrbuch, Systematik*, **106**, 406–26.

Silvers, W. K. (1979). *The Coat Colors in Mice*. New York, Heidelberg, Berlin: Springer-Verlag.

Selected references for Chapter 9

Apfelbach, R. & Ebel, K. (1975). Vom Suchbildverhalten des Frettchens (*Putorius furo*) beim Beutefang. *Zeitschrift für Säugetierkunde*, **40**, 378–9.

Curio, E. (1976). *The Ethology of Predation*. Berlin, Heidelberg, New York: Springer-Verlag.

Gilmore, R. M. (1963). Fauna and ethnozoology of South America. In *Hand-*

book of South American Indians, vol. 6, ed. J. H. Steward, pp. 345–464, New York: Cooper Sq. Pubs Inc.

Todd, N. B. (1977). Cats and commerce. *Scientific American*, **237**, 100–07.

Selected references for Chapter 10

Buttersworth, M. H., Steinhauf, D. & Weniger, J. H. (1965). Streβresistenz als Leistungsmerkmal beim Schwein. 2. Mitteilung: Methoden zur Bestimmung der Streβanfälligkeit und erste Ergebnisse über Beziehungen zwischen Streβanfälligkeit und Leistungsmerkmalen beim Schwein. *Züchtungskunde*, **39**, 283–94.

Klatt, G. & Schlisske, W. (1974). Einflüsse der bewegungsarmen Haltung gravider Sauen bei extrem verkürzter Säugezeit auf die Leistung. *Archiv für Tierzucht*, **17**, 287–98.

Krech, D., Rosenzweig, M. & Bennet, E. L. (1962). Relations between brain chemistry and problem-solving among rats raised in enriched and impoverished environments. *Journal of Comparative Physiological Psychology*, **55**, 801–07.

Selected references for Chapter 11

Bree, P. J. H. van, Soest, R. W. M. van & Vetter, J. C. M. (1970). Biometric analysis of the effect of castration on the skull of the male domestic cat (*Felis catus* L., 1958). *Publicaties van het Natuurhistorisch Genootschap in Limburg*, **20** (3/4), 11–14.

Derenne, P. (1972). Donnés craniometriques sur le chat haret (*Felis catus*) de l'archipel de Kerguelen. *Mammalia*, **36**, 459–81.

Hückinghaus, F. (1965). Craniometrische Untersuchung an verwilderten Hauskaninchen von den Kerguelen. *Zeitschrift für wissenschaftliche Zoologie*, **171**, 183–96.

Kruska, D. & Röhrs, M. (1974). Comparative-quantitative investigations on brains of feral pigs from the Galapagos islands and of European domestic pigs. *Zeitschrift für Anatomie und Entwicklungsgeschichte*, **144**, 61–73.

Menzel, R. & Menzel, R. (1960). *Pariahunde*. Wittenberg: Ziemsen.

Selected references for Chapter 12

Belyaev, D. K. & Trut, L. N. (1975). Some genetic and endocrine effects of selection for domestication in silver foxes. In *The Wild Canids*, ed. M. W. Fox, pp. 416–26. New York, Cincinnati, Toronto, Melbourne: Van Nostrand Reinhold.

Blaxter, K. L., Kay, R. N. B., Sharman, G. A. M., Cunningham, J. M. M. & Hamilton, W. J. (1974). *Farming the Red Deer*. Edinburgh: HMSO.

Eich, E., Hemmer, H. & Reichert, E. (1979). Studien zur Ansatzmöglichkeit einer Domestikation des Steppenzebras, *Equus* (*Hippotigris*) *quagga* Gmelin, 1788. *Säugetierkundliche Mitteilungen*, **27**, 147–56.

Hemmer, H. (ed.) (1986). *Nutztier Damhirsch*. Bonn: Rheinischer Landwirtschaftsverlag.

Heptner, W. G. & Nasimowitxch, A. A. (1967). *Der Elch*. Wittenberg: Ziemsen.

Reinken, G. (1980). *Damtierhaltung auf Grün- und Brachland*. Stuttgart: Ulmer.

Treus, V. D. & Lobanov, N. V. (1971). Acclimatisation and domestication of the eland *Taurotragus oryx* at Askanya-Nova Zoo. *International Zoo Yearbook*, **11**, 147–56.

Selected references for Chapter 13

Bohlken, H. (1961). Haustiere und zoologische Systematik. *Zeitschrift für Tierzüchtung und Züchtungsbiologie*, **76**, 107–13.

Cohen, M. N. (1977). *The Food Crisis in Prehistory*. New Haven, CN; London: Yale University Press.

Dennler de la Tour, G. (1968). Zur Frage der Haustier-Nomenklatur. *Säugetierkundliche Mitteilungen*, **16**, 1–20.

Groves, C. P. (1971). Request for a declaration modifying Article 1 so as to exclude names proposed for domestic animals from zoological nomenclature. *Bulletin of Zoological Nomenclature*, **27**, 269–72.

Reimov, R., Adamczyk, C. & Andrzejewski, R. (1968). Some indices of the behaviour of wild and laboratory house mice in a mixed population. *Acta theriologica*, **13**, 129–50.

Stodart, E. & Myers, K. (1964). A comparison of behaviour, reproduction, and mortality of wild and domestic rabbits in confined populations. *Commonwealth Scientific and Industrial Research Organization Wildlife Research*, **9**, 144–59.

Tomich, P. Q (1969). *Mammals in Hawaii*. Honolulu: Bishop Museum.

List of photographs taken by the author in public zoos and animal parks

Place	Figure number
1. Zoos and animal parks	
Berlin, Germany	3.14
Dortmund, F.R.G.	3.16, 3.30, 9.15, 9.17, 12.3, 12.4, 12.5, 12.6
Duisberg, F.R.G.	2.11, 12.7
Frankfurt, F.R.G .	3.15, 3.28
Hannover, F.R.G .	3.2
Honolulu, U.S.A.	3.43
Cologne, F.R.G.	2.24, 3.1
Kronberg, F.R.G.	9.12, 12.2
Munich, F.R.G.	2.12, 2.23, 3.18, 3.20, 3.21,3.26, 3.29, 3.31, 3.38, 3.39, 3.40, 9.18, 9.19, 12.11
Münster, F.R.G.	2.21
Rabat-Témara, Morocco	2.20, 9.16
Rotterdam, Netherlands	3.32
San Francisco, U.S.A.	3.27
Vienna, Austria	9.11
2. Game parks, small zoos	
Birkenfeld, F.R.G.	3.37
Chillingham Castle, England	9.2
Darmstadt, F.R.G.	9.1
Hilo, Hawaii, U.S.A.	9.14
Kaiserslautern F.R.G.	3.34
Mainz-Gonsenheim, F.R.G.	3.36, 3.42, 4.1, 9.22
Rheinböllen, F.R.G.	12.12
Wiesbaden, F.R.G.	2.4, 12.9

Index

Italic page numbers denote figures.